AN ANTHROPOGENIC TABLE OF ELEMENTS

Experiments in the Fundamental

Edited by Timothy Neale, Courtney Addison, and Thao Phan

An Anthropogenic Table of Elements provides a contemporary rethinking of Dmitri Mendeleev's periodic table of elements, bringing together "elemental" stories to reflect on everyday life in the Anthropocene.

Concise and engaging, this book provides stories of scale, toxicity, and temporality that extrapolate on ideas surrounding ethics, politics, and materiality that are fundamental to this contemporary moment. Examining elemental objects and forces, including carbon, mould, cheese, ice, and viruses, the contributors question what elemental forms are still waiting to emerge and what political possibilities of justice and environmental reparation they might usher into the world.

Bringing together anthropologists, historians, and media studies scholars, this book tests a range of possible ways to tabulate and narrate the elemental as a way to bring into view fresh discussion on material constitutions and, thereby, new ethical stances, responsibilities, and power relations. In doing so, *An Anthropogenic Table of Elements* demonstrates through elementality that even the smallest and humblest stories are capable of powerful effects and vast journeys across time and space.

(Technoscience and Society)

TIMOTHY NEALE is a DECRA senior research fellow and senior lecturer in anthropology at Deakin University.

COURTNEY ADDISON is a lecturer in the Centre for Science in Society at Te Herenga Waka, Victoria University of Wellington.

THAO PHAN is a research fellow at the ARC Centre of Excellence for Automated Decision-Making and Society and the Emerging Technologies Research Lab at Monash University.

TECHNOSCIENCE & SOCIETY

If our world and our futures are technoscientific, then how should we organize this world? And how should we understand these futures? Technoscience and Society seeks to provide new analytical tools to do this, as well as new empirical insights into the changes happening around us. The series encourages shorter, punchier scholarly books providing a crossover forum in which both established researchers and new and emerging scholars can present their investigations into the ever-changing relationship between technoscience and society.

AN ANTHROPOGENIC TABLE OF ELEMENTS

experiments in the fundamental

edited by
timothy neale
courtney addison
thao phan

UNIVERSITY OF TORONTO PRESS
Toronto Buffalo London

© University of Toronto Press 2022
Toronto Buffalo London
utorontopress.com
Printed in the U.S.A.

ISBN 978-1-4875-6356-1 (cloth)   ISBN 978-1-4875-6359-2 (EPUB)
ISBN 978-1-4875-6357-8 (paper)   ISBN 978-1-4875-6358-5 (PDF)

Technoscience and Society

Library and Archives Canada Cataloguing in Publication

Title: An anthropogenic table of elements : experiments in the fundamental /
edited by Timothy Neale, Courtney Addison, and Thao Phan.
Names: Neale, Timothy, editor. | Addison, Courtney, editor. | Phan, Thao, editor.
Description: Series statement: Technoscience and society | Includes
bibliographical references and index.
Identifiers: Canadiana (print) 20220243913 | Canadiana (ebook) 2022024409X |
ISBN 9781487563578 (paper) | ISBN 9781487563561 (cloth) |
ISBN 9781487563592 (EPUB) | ISBN 9781487563585 (PDF)
Subjects: LCSH: Nature – Effect of human beings on. | LCSH: Periodic table of
the elements.
Classification: LCC GF75 .A58 2022 | DDC 304.2 – dc23

Cover image credits appear on page 239.

We wish to acknowledge the land on which the University of Toronto Press
operates. This land is the traditional territory of the Wendat, the Anishnaabeg,
the Haudenosaunee, the Métis, and the Mississaugas of the Credit First Nation.

This book has been supported by a grant from the Alfred Deakin Institute for
Citizenship and Globalisation at Deakin University, Australia.

University of Toronto Press acknowledges the financial support of the
Government of Canada, the Canada Council for the Arts, and the Ontario Arts
Council, an agency of the Government of Ontario, for its publishing activities.

Canada Council   Conseil des Arts
for the Arts   du Canada

ONTARIO ARTS COUNCIL
CONSEIL DES ARTS DE L'ONTARIO
an Ontario government agency
un organisme du gouvernement de l'Ontario

Funded by the   Financé par le
Government   gouvernement
of Canada   du Canada

Canadä

CONTENTS

AN ANTHROPOGENIC TABLE OF ELEMENTS

INTRODUCTION

timothy neale
courtney addison
thao phan

Dmitri Mendeleev's Periodic Table of Chemical Elements is one of the most iconic and enduring symbols of scientific thought. In its tidy arrangement of substances, identities, atomic weights, and other properties, it illustrates a central project at the heart of the chemical sciences: to understand the universal structures that define life's elementary matter.[1] In their 2019 statement commemorating the Table's 150th anniversary, the editors of the journal *Nature* remarked on its worldwide appeal, lauding its ability to communicate across varied audiences and its prominent status not just within science but also in the broader cultural imagination. Significantly, they also emphasized the Table's utility in changing our perceptions of the universe from something vast and unknowable to something finite and tractable. "There is clearly something about the [Periodic Table] that resonates with a wider audience," they wrote. "Chemists should seek to tap into this fascination in the year ahead and highlight the importance of the original and still the best – the one that corrals all of the known atomic building blocks of the universe into an orderly array."[2] While Mendeleev was neither the first nor the last person to chart the chemical elements, it was this instrumentalist promise to give meaning, structure, and order to an otherwise chaotic world that gave *this* table its longevity.

This book revisits the Periodic Table and its promise to make the elemental knowable. In its earliest iterations, the items mapped in the Table were said to represent a distinct material substance, something that has "yet to be broken down into any more fundamental components by chemical means."[3] This conceptualization of the essential, material order of things encourages us to see the world as a derivative or composite of chemicals and their molecular realities, rather than as a mess of complicated stories and structures that cannot be reduced to their constituent parts. At the same time, Mendeleev's foresight to hold open spaces for elements that had not yet been discovered in 1869 suggests a humility and optimism towards the future – an awareness that the world can "kick back" (in its own time) and a reminder, as Addison states, "that what mattered in 1869 is not necessarily what matters now."[4] Though the Periodic Table has changed significantly over its lifetime – there have been over 700 versions since Mendeleev's publication alone – to know and map life's matter remains a captivating premise for many within and outside the chemical sciences. The authors in this book have each taken the anniversary of the Periodic Table as an invitation to participate in a chain of collaborative experiments with the elemental and reflect on the very notion of elementality.

For the editors, these ideas first bubbled to the surface during the Anthropocene Campus Melbourne in 2018, an event where many of the authors first met and gathered to play with earth, fire, water and air, also known as the "classical elements" due to their common centrality to several ancient cosmologies.[5] These familiar entities were made strange through workshops and field trips, encountering designed future fire ecologies, air engineered for human consumption, waters as sites of more-than-human obligation, and earthen mounds of nonbiodegradable metal-heavy human excrement that will likely outlast human life.[6] Inspired by these uncanny elements and provoked by Mendeleev's chart, we developed a series of essays seeking to "better represent the constitutive elements of our current moment" through short narratives about the curious social lives of constitutive companions, such as aquifers, clouds, micronutrients, and radioactive isotopes.[7]

An experiment at the 2019 meeting of the Society for the Social Studies of Science in New Orleans followed, with an invitation for conference attendees to sit with us at a physical dinner table.[8] In the centre of the table, covered with a white tablecloth, were plates arranged with cards that had printed on them the names of different elemental entities – concrete, glass, testosterone, and many others. Each guest was given a "menu" asking them to speculate on the precursors, co-existences, and afterlives of their element (see figure 0.1). With only a little prompting, our guests formulated impromptu "implosions" locating these abstractions in specific materialities, plucking at their "sticky economic, technical, political, organic, historical, mythic, and textual threads."[9] As at the best meals, our guests sometimes became unruly, talking over one another, raising questions about unseen connections, obscure histories, or cryptic resonances between one ringing thread and another.[10]

Like the iterative drafts of Mendeleev and others, our experiments with the elemental have built on each other, while also making it increasingly clear that any selection or account of the elemental is necessarily

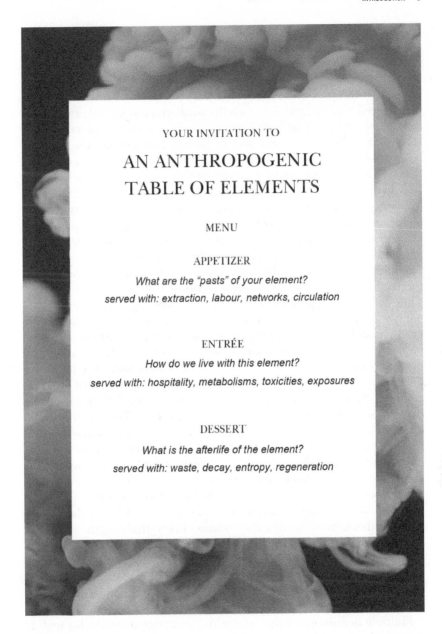

Figure 0.1. Invitation to *An Anthropogenic Table of Elements* (credit: Timothy Neale, Courtney Addison, and Thao Phan)

partial. It is this sense of partiality – a self-reflexive situatedness – that we hope to accentuate through each iterative test of possible ways to tabulate and narrate the elemental. While phrases such as "the elemental" and "the Anthropocene" gesture to vast scales anchored in deep time and impossibly complex and interconnected planetary narratives, the chapters arrayed here suggest that even the smallest, most humble stories are capable of powerful effects and vast interscalar journeys.[11] To start a story close to the subject, close at hand, or close to home (wherever that may be) invites engagement on more perceptible scales – or at least, on scales that matter. Partial stories by their nature resist any expectation that they could be mistaken for the whole. They lay bare that "incompleteness is inherent," as Rider argues in her chapter, and therefore that any table or collection of the elemental could ever possibly be complete for all of us.[12]

Small stories also have enough space to name each actor. Some actors demand a lot of space – capitalism, settler colonialism, extractivism, exploitation, for example – but in naming these actors, it can be easy to forget that each of these is in turn composed of innumerable others. These others often struggle to find space in the big stories we typically tell, yet they become the centre of every small story: the Chilean government and its embrace of copper extraction, the mining corporations piercing Albertan soil, the long strontium shadow of the US military's nuclear testing. In one experiment at the Haus der Kulturen der Welt's "The Shape of a Practice" event in 2020, education scholar Denise Frazier used the framework we had developed to map the "elements" of the carceral state in New Orleans and the wider US. The Anthropocene, disaster capitalism, racism, and pandemics make up one temporal series, while the 32 elements at the centre of the table name some of the people who were killed through police violence in the United States during the past several years (see figure 0.2). Other rows specify elements relating to social movements, histories of redlining and removal, fenceline communities exposed to industrial toxins, and the Atlantic slave trade among others. In such ways, the contributors to this volume ask: What is elemental to this moment? What elemental forms are yet to emerge? And what political possibilities of justice and environmental reparation might they usher in? One proposition of this book is that experimenting with alternative ways of representing elemental forms might generate new possibilities to help address these questions.

Multiple definitions of the "elemental" circulate through this text. It is worth clarifying some of their features so you will know them on sight, sneaking through a subclause, or peeking out of a footnote. For our purposes, the first meaning is the philosophical proposition that existence is structured by certain things – atoms, humours, spirits, energies, states

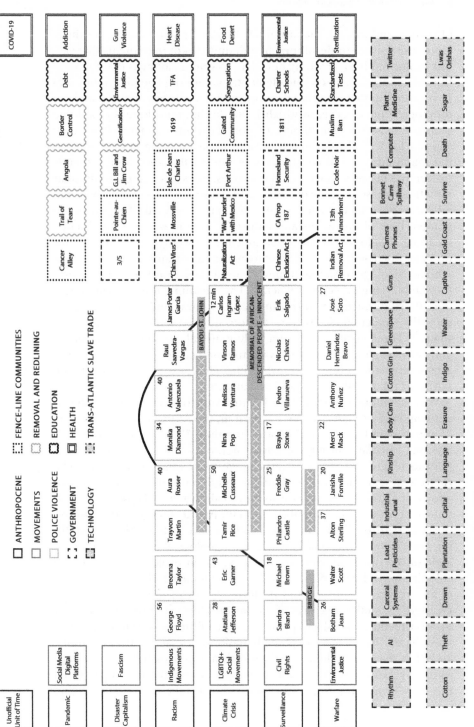

Figure 0.2. An anthropogenic table of elements (credit: Denise Frazier)

of matter, and so on – that are more fundamental or vital than others. Here, existence is underpinned by a definite set of ingredients that cannot be derived from anything else. The second meaning is that found in the Periodic Table, where the elemental names the chemical molecules or compounds that compose the universe's constitutive matter. This is an ontological perspective, in which finite entities constitute the building blocks of existence. Put bluntly, carbon is carbon no matter how much you pay for it.[13] Third, the infrastructural meaning figures elements as the manifold material constituents of our Earthly surrounds. Organic matter and processes such as soil microbes, trees, oceanic currents, solar radiation, and subterranean hydrocarbons are the basis of our elemental milieu or what media scholar John Durham Peters calls "the taken-for-granted base of our habits and habitat."[14] Finally, we have the situated meaning developed in this book and introduced by the experiments we have discussed above. "Elements never fully stand alone," as media scholar Nicole Starosielski argues. "They attach, bond, and transform"; they "mix, mingle, blend and act."[15] This meaning, thereby, cuts across the others, diffracting and reading between these multiple meanings to show the elemental as always in relation, situated somewhere and some-when.[16] These relations are not only material: in some cases (where authors have, for example, analysed the molecular contents of ice, and artists project meaning onto mylar film) elements become tools themselves for reading the world and one another.[17]

Why elemental thinking in or for the Anthropocene, specifically? As anthropologist Cymene Howe has written, the Anthropocene as "a state of impairment has a way of focusing attentions."[18] Interpretations of this age that circulate through scientific networks and social media feeds tend to regard this era as a step change from what came before. The rhetoric of extreme, unprecedented change certainly seems to provoke a form of alarm that could morph into a call to action. However, in their critical account of the scientific and cultural debates over the Anthropocene, Heather Davis and Zoe Todd remind us that this temporalizing endeavour of bookmarking a human geological epoch "continues a logic of the universal" that was instrumental in projects of empire and colonization.[19] Like the Periodic Table of Elements, the Anthropocene as an organizing concept and performative descriptor is a product of (Western) scientific readings of the world. There are myriad other ways to describe the epochal and planetary scale of the mess we find ourselves in – climate change, ecocide, the Plantationocene, the sixth extinction – and while each carry their own diagnostic advantages and baggage, only some of them point to the vast disparities in benefits and impacts that this mess brings. But what unites these terms is the understanding that more-than-human survival is now in question, pressing humans to examine the fundamental bases of damaged and diminishing ecological envelopes for existence. If the elemental bespeaks endurance, and the Anthropocene signifies massive scale, then their critical conjunction perhaps attunes us to traffic across scales, and to questions of what lasts and with what effects.

It is the play between these foundational and relational perspectives that we hope can offer some intellectual purchase for the present moment. Where thinking elementally attunes us to foundational forces and constituents, the Anthropocene foregrounds the protracted relations those forces form and distend. Thus, we see shifting tectonic plates become the literal ground for geopolitical manoeuvring and guerrilla resistance in Turkey,[20] and the constant biological turnover of sperm perturbed by both political anxieties and industrial pollution in China.[21] Thinking elementally also offers a much-needed humility for addressing the Anthropocene. The elemental may indeed exceed human capacity to grasp it,[22] but it also subverts. "To engage with the Anthropocene," as Hannah Gibson and Sita Venkatsawar remind us, "is also an invitation to dismantle the rifts that separate humans from other life forms."[23]

While for some the elemental offers escape from the human, this book is grounded in a sense that there is an urgent need to always keep humans implicated and within the frame of analysis. We attach the constraint of "anthropogenic" to our experiments not to *centre* peoples' thoughts and actions – their power is arguably minor, unintentional or tangential in many contexts – but rather to insist they stay in the mix. Indeed, as each chapter demonstrates, the human is inextricably situated in networks of animal, chemical, and other material relations. Whereas Mendeleev's table attempts to chart laws of a stable universal order, the abundant evidence that we have slipped from stability into entropy (and possible annihilation) calls for a renewed attention to the infrastructure of contemporary anthropoid existence, "teeming with time and history," that has been naturalized into anonymity.[24]

Thinking elementally, as geographers Sasha Engelman and Derek McCormack note, "reminds us of how forms of life live, move, and associate by virtue of their immersion" in and with one another in different milieus, shaped by continuing histories of racial, gendered, and economic injustice.[25] The chapters in this volume illustrate these injustices in different ways and at different scales: through the strontium leached in bodies from early nuclear experiments in the US; the sodium fluoroacetate (the compound 1080) used to promote life through selective death in Aotearoa New Zealand; the nationalistic and geopolitical drive that promoted cement as a modernizing force in Thailand; the pharmaceutical use of the toxic chemical lithium as a mood stabilizer in the body; the use of testosterone as a measure for sporting participation and rights, and many more.[26]

We have little interest in taking flight from human involvement, not only because the crises of interest are here on Earth (and places closer still) but also because human influence and impact has extended in every direction. The echo of human impacts can be followed upwards beyond the exosphere, into interstellar space, and downwards into the lithosphere, through holes up to 12 kilometres deep. These holes, which cut through the geological record of the hundreds of thousands of years of *Homo sapiens'* presence, have been drilled to access the combustible oil and gas formed by ancient sunlight to fuel forms of industrial life whose radioactive and toxic aftermath will likely remain legible on this planet until it is engulfed by our sun in a few billion years.[27] Of course, writing in this way pushes attention to the outer spatial and temporal limits of the anthropogenic, when our intent is to insist that it is we humans who might begin the work of responding and bringing about less damaged and damaging ways of living. As the anthropologist Gabrielle Hecht argues, one starting point might be "interscalar vehicles" that can take us between the atomic and the atmospheric, and other scales besides, crafting emplaced narrations while also "keeping the planet and all of its humans in the same conceptual frame."[28] Like others before her, Hecht seeks to draw attention to "who pays the price for humanity's planetary footprints," but by attending to scalar boundaries and, thereby, the histories, structures, and interested actors who work to maintain those boundaries, holding networks and worlds apart and holding inequities in position.

How, then, to go about this work of mapping and analysing connections across spatiotemporal scales and contexts? This book draws inspiration from the field of Feminist Science and Technology Studies (STS), particularly from those writers who have sought to explore how chemical elements have been integrated into the reproduction or endangerment of human and more-than-human life. Historian and anthropologist of science Michelle Murphy, whose works reverberate through many chapters in this book, has critiqued how research and governance institutions conceptualize chemicals as discrete entities and thereby, through their practices, co-constitute them as such. One by one, chemicals are named, described, and inserted into regulatory frameworks and agreements. One by one, these entities are attached to biophysical effects and ailments, depending on the weight of evidence connecting that entity to a certain effect in an avowedly representative population. In this way, the elemental thinking captured by convoluted names such as

Dichlorodiphenyltrichloroethane (DDT) or Polychlorinated biphenyls (PCBs) "pushes aside [their] complex reactivity with living- and non-living-being," obscuring the complexity and inequity of our respective exposures to toxicity and, as several chapters detail, longer histories of dispossession and colonization.[29] The translocal dominance of this discourse means that it is hard not to talk about chemicals in such universalizing and abstract terms. If we are to have any hope of undermining the molecularization of harm – and its disproportionate effects on racialized bodies – we need to find new ways to track and narrate "the infrastructure of chemical relations that surround and make us."[30]

Murphy and their collaborators are not alone in calling for (or demonstrating) new narratives that track elemental relations.[31] Sociologist Hannah Landecker has spoken about "the Anthropocene of the cell" as a critique that articulates "a historically unfolding biology of this world."[32] While the Anthropocene is often conceived in terms of climatic causes with downstream biological effects, Landecker notes that the rise of petrochemical industrialism has also created pathologies, such as arsenicals and antibiotic resistance, that register at a cellular level. Exposed human and nonhuman metabolisms must be understood, following Tironi, as part of "the extended ecology of industrial extraction."[33] Attention to such metabolic relations has the potential to act "like a biological stain," Landecker suggests, throwing into relief structures, boundaries, and vectors that are otherwise faint or imperceptible.[34] To do this necessarily involves the repudiation of what social and political theorist Alexis Shotwell calls "purity politics." Being pure, or untouched, is a practical impossibility in a world of messy entanglement, and so Shotwell insists that we "see complexity and complicity as the constitutive situation of our lives."[35] Analysing this complexity, she argues, will reveal how we are collectively implicated in "distributed mortality" across species lines. As Chako and Neale argue in their chapters, elemental logics are necessary to the commensuration, commodification, and exploitation of entities such as plant seeds and carbon emissions, and so in approaching new and old elemental grammars we must remain alert that they, like written languages, have their politics and consequences.[36]

Alongside these scholars, and others besides, the work of philosopher Karen Barad has provided significant guidance regarding the hazards and rewards of elemental thinking and interscalar navigation. Barad's extensive writing revolves around the crafting and use of analogical portals between quantum, atomic, interpersonal, and geological realms in order to reflect on the queerness of existence itself. They have argued, contrary to conceptualizations of matter as something that might be contained or put out of relation – entombed from others' interference – that radical openness or hospitality is written into matter itself.[37] To bring to light the relational tangles and muddles of a changing and contingent world, Barad suggests, we must build "diffraction apparatuses" or analytical instruments that read the insights of different scales and realms through one another.[38] Hecht, Murphy, Landecker, and Shotwell all differ significantly in their approaches from Barad; however, we find an inspiration to our approach to the elemental in their shared resolve that there is always room to travel further still; to venture outside the typical temporal and spatial envelopes of human life to track, and tack between, itineraries of the elemental.[39] In an elementally entangled world, as Wark concludes in his chapter, the best way forward "is to join their circulations and to find what they enfold."[40]

Though a focus of the elemental aligns closely with recent growing interest in our chemical and infrastructural milieu, elemental thought is also arguably out of kilter with other contemporary analytical instincts. While many are now confronting the possibility of human extinction at a planetary scale, we should also remember that the allure of the elemental is that it compels us to think beyond the human. The elemental, Engelman and McCormack note, "both captures something tangible about the world and also remains excessive of human agency or intervention."[41] Like the concept of the Anthropocene, it "pulls in contrary directions," drawing our attention both to the basis of human life and "what has definite and forceful existence regardless of our sense of world," or even our very presence.[42] Humans are not essential to existence when viewed from the inhuman perspective of the Universe. Indeed, while this volume highlights the many damages wrought by human actors and their techno-chemical collaborators, it also hints to the afterlives of these nonhuman entities. In some form or another the world's ice sheets, tectonic plates, and atmospheric molecules will persist irrespective of our presence. Uranium does not need our help, let alone our presence, to leak its electrons. Dark matter could care less if we are here or not. Anthropogenic influences may pervade the Earth's biosphere and atmosphere, producing traces of industrial farming and extractivism that etch human presence into the lithosphere, but one should not presume to know what the planet is thinking. Earth pre-existed our emergence and would remain if all of us suddenly, magically, exited.

There is another elemental tension worth confronting. The arrival of the Anthropocene concept has further buoyed the established influence of certain critical thinkers – Deleuze, Haraway, Braidotti, and Barad, to name a few – whose unifying themes are emergence and entanglement; all beings are, in Haraway's words, "intra-active entities-in-assemblages."[43]

Scholars "attuning to multispecies entanglement" search for hauntings, ghosts, and "the liveliness of landscapes," documenting and theorizing what anthropologists João Biehl and Peter Locke describe as "transformative visions and potentials emerging from unexpected corners."[44] As anthropologist Anna Tsing proposes, "precarity is the condition of our time," leading to a framing of social worlds as contingent struggles within a "polyphonic assemblage ... [of] open-ended gatherings" that we might otherwise call life. To therefore contend, like some latter-day Aristotle, that all matter is the determinate expression of a handful of elements seems antiquated (even regressive). If all being is built out of a set of entities so concise that they can be enumerated on one hand, or spoken in one breath, then existence would seem to be less "polyphonic" than binary or contrapuntal.

However, the authors in this book draw upon these theorists to contend that elemental thinking is not only compatible with theories of emergence and polyphony, but constitutive of them – and the more potent for it. They claim elementality for a diverse set of objects and use to track the mess that emerges from seemingly orderly materials. The historian of science Eric Scerri notes that Mendeleev himself harboured strong Aristotelian leanings, seeking out the "simple substances" and simple order that underpinned the superficial diversity of material forms.[45] What links these and other examples in the array of elemental thinking is their ontological premise that a syntax or grammar of certain indispensable things underpins everything. A table of the elemental is thereby a lexicon for all being or, to use Cohen and Duckert's phrase, "matter's kinetic syllabary."[46] An elemental perspective is one that tries to locate an analytic starting point and then trace the cacophony of reactions that unfold from there. It is this perspective, which conjoins a finite number of elements with innumerable recombinant possibilities, that can both explain the world and offer us ways of better acting within it. It is an opportunity to push back and repurpose the conceit of universality for our own ends.

A claim to the elemental is therefore both methodological and political. Methodologically, it indicates the need for an analytic starting point and seeks to find one. It is also a political claim to what is fundamental and therefore worthy of our attention. Indeed, what is fundamental to the present moment is not entirely the same as what was fundamental to Mendeleev and his peers – or even, fundamental to their descendants in the chemical sciences. The elemental, in other words, might be mobilized to insist on the pervasive and persistent influence of certain human histories – histories of, for example, capitalist extractivism, colonialism, and settler-colonialism. The material traces of empires exercise force, as historian Ann Laura Stoler has argued, even as they become ruins and rubble for subsequent "postcolonial" social worlds.[47] Or, as scholars such as Alex Weheliye and Sylvia Wynter have contended, race and racialized figurations remain foundational (or elemental) to contemporary categories of the human and non-human alike.[48] Working in an elemental mode, we may agree on a basic physical structure of the world, but acknowledge that the matters of most pressing concern and their material constitution have changed and, with them, our ethical quandaries.[49] In this sense, the chapters in this book can be considered performative experiments in elemental thinking. They are inquiries that mobilize the ontological precept of elemental philosophies to bring into view new material constitutions and, thereby, new ethical stances, responsibilities, and power relations.[50]

The chapters that follow take one of two approaches: some trace how "traditional" chemical elements (such as carbon, lithium, strontium) have moved and morphed across time and place, revealing their complicity in anything but natural circuits of power; some apply the label of the elemental to different entities entirely (for example, sperm, seeds, cement), making the case that objects that may seem ancillary are actually also the germ of other politics and material realities. What does it do to these various objects and substances if we term them elemental to a situation or predicament? Or, put differently, what does elemental thinking do for us? Exactly which things are presented or chosen as foundational, and why – these are precisely the kinds of difficult questions that our own elemental thinking seeks to impress. A certain ignorance is, Chako states, "a requisite feature of maintaining the illusion of elementality."[51] Each claim therefore reflects a set of situated choices that exposes the irreducible role of subjectivity in naming and describing our relations to and with the world. This reminds us that we are not outside of elementary matter but are a part of it: constituting it as it constitutes us.

To term something "elemental" is thereby, necessarily, both a political and deeply intimate claim to importance at the level of being. It is an acknowledgment that our own bodies and subjectivities are a part of the scientific apparatus of knowing.[52] This, for instance, is illuminated by Wark, who charts how lithium's traffic enfolds bodies and worlds, and by Latham and Seear in their evocative account of how testosterone and human rights enjoin one another. Claims to the elementality of different entities and collections of entities other than those found in the Periodic Table can therefore be understood as interventions, decentring scientific thinking or even refusing the persistent imbalance between scientific and other ways of knowing the affinities and bonds that make up the world. With newfound perspective on elemental injustices we might be

able to explore and craft new modes of intervention or name and mobilize as-yet unnamed collectives of concern.

The powerful and creative uses of the elemental we have surveyed
are not only a reminder of the philosophical and political interventions
such thinking can make but also that, inevitably, there are many more
genealogies of elemental thought than those that feature prominently
here. Han-era (206 BCE–220 CE) *wu xing* 五行 models of nature that
are central to contemporary traditional Chinese medicine, for example, foreground the relationship between the five elemental movements
(water, fire, wood, metal, and earth). Many different Aboriginal laws and
lores on the continent now known as Australia, to give another example,
figure fire as a force that pre-exists humans and comes to them through
the benevolent or imprudent action of the nonhuman creators of existence.[53] In this vein, some of the authors here highlight the *logic* of
elementality that contoured Mendeleev's own thought, but also characterized anthropology's structuralist period, and persists today in carbon
markets, and seed banking enterprises.[54] In short, as the chapters that
follow demonstrate, the elemental has not just been the preoccupation
of Socratic philosophers, Western humanities scholars, and Russian
chemists alone. This book is a speculative table, an open-ended rendering of our predicament in the form of ethnographically-grounded
stories, assembled to demonstrate the possibilities and importance of
different iterations of the elemental. These are infrastructural stories
that sketch always unfinished ontologies – stories, recalling Shotwell's
words, of "complexity and complicity" to denaturalize our naturalized
habitus, stories of how we humans in our more-than-human worlds are
never, really, out of our element.

In Mendeleev's Periodic Table, the arrangements of elements can be
read in a number of ways. Reading left to right, top to bottom, they are
arranged by their atomic weight. But then there are also columns – the
alkali metals and the noble gases, for example – and rows – the lantha-
nides and actinides – of related members, as well as the ziggurat of metal-
loids and wedge of reactive non-metals. Similarly, the chapters that follow
can be arranged and navigated by readers in a number of ways. Where
Mendeleev's Table is ordered according to atomic weights, electron con-
figuration, and other properties, here we have surrendered some of our
prescriptive power as editors over these strange and multiple stories and
presented them in alphabetical order. To take the conceit of the Periodic
Table and subject it to the arbitrariness of the Latin alphabet is to play
with the affordances of structure and disarray, to invite illogical connections that nonetheless attach and resonate. This counter-visualization

appropriates one dominant scientific rendering of the world and attempts to retrofit it for the Anthropocene.[55] Though not "periodic," the chapters here collectively chart the character and relationship between elements to emphasize recurring themes and common properties across media that appear at first glance to be different things entirely.

As noted above, this book has been compiled less as a definitive collection, and more as an open-ended catalogue. They represent just some of the elements we have managed to assemble so far and, as Phan's chapter argues, gesture to the ones "we do not yet know but are compelled to keep space for nonetheless."[56] Rather than a linear table of contents – and in the spirit of Mendeleev's original formulation – we instead present the chapters in a tabularized chart. In what follows (see figure 0.3), one could read according to entity type, tracking between chemical elements (carbon, copper, lithium, strontium), organic life-forms (cheese, mould, seed, virus), material strata (cement, kerosphere, tectonics, ice), synthetic compounds (1080 or sodium monofluoroacetate, mylar), bodily substances (sperm, testosterone), and as-yet undiscovered or speculative matter (elements-to-come). Another pathway is to read for geography, beginning your itinerary in Australasia (1080, carbon, cement, testosterone), Turtle Island (ice, kerosphere, mould, mylar, strontium), or elsewhere (copper, lithium, sperm, virus). Further readers might wish to negotiate the text through theoretical formations, and their borderlands, perhaps starting with those closest to Mendeleev (e.g., cheese), the Hecht (e.g., carbon, cement) and Murphy clusters (e.g., 1080, mylar), or tacking between subbranches of the Barad (e.g., kerosphere, elements-to-come) and Haraway lines (e.g., seeds, virus).

Finally, readers might encounter the chapters in a random or alphabetical order and nonetheless be struck by the frequency with which certain narratives of capitalism, settler-colonialism, militarism, racism, modernization, and extractivism return as definitional to the Anthropocene across scale. Engaging with destructive systems of production and more-than-human others may now be a convention of "Anthropocene writing" as genre, but taken together, the chapters of this book demonstrate just how many permutations these forces take on across different contexts. Consistently, we see single elemental forms emerge in military research projects and find their way into domestic spaces or faraway fleshy bodies. Efforts to count and corral other elements become central to corresponding efforts to classify and control subjects and populations. In the chapters here, geological and historical time collapse into each other – into untold futures – while molecular jitters reverberate across landscape, strata, and ecosystem. Sciences (geoscience, reproductive biology, toxicology, botany, and so on) are repeatedly called upon to

Figure 0.3. A possible table of contents for *An Anthropogenic Table of Elements* (credit: Emma Holder)

interpret the signs thrown up by the Anthropocene, to articulate the stakes of what is unfolding, and issue predictions from necessarily imperfect knowledge. The nation-state as arbiter of life, risk, and rights, is pervasive.

Thinking elementally, as we have discussed, brings renewed focus to the infrastructural and ontological possibilities of this state of impairment, renewed attention to how futures have been foreclosed and what might yet be composed out of what remains. In charting these relations, and the traffic of material entities, be they molecular, geologic, or something else entirely, across domestic, industrial, and nonhuman settings, the chapters in this book enjoin us to create common ground out of uncommon elements. Like Mendeleev's Periodic Table, we hope that this book serves as both a guide to reactivity and a resource for its understanding, mapping those elements most susceptible to transformation, plotting sites where new entities might emerge or announce themselves. The indivisibility of the elements from the political and moral questions of the Anthropocene and the potential for elemental thinking to catalyse more just and sustainable futures is precisely what this collection seeks to impress.

NOTES

1 In Mendeleev's own textbook he defined "chemistry" as "a natural science which describes homogeneous bodies, studies the molecular phenomena by which these bodies undergo transformations into new homogeneous bodies, and as an exact science it strives ... to attribute weight and measure to all bodies and phenomena, and to recognize the exact numerical laws which govern the variety of its studied forms." See Mendeleev, in Michael Gordin, *A Well-Ordered Thing: Dmitrii Mendeleev and the Shadow of the Periodic Table*, rev. ed. (Princeton University Press, 2019), 22.

2 "Anniversary Celebrations Are Due for Mendeleev's Periodic Table," *Nature* 565, no. 7741 (2019).

3 Eric R. Scerri, *The Periodic Table: Its Story and Its Significance* (New York, NY: Oxford University Press, 2007).

4 Karen Barad, *Meeting the Universe Halfway: Quantum Physics and the Entanglement of Matter and Meaning* (Durham, NC: Duke University Press, 2007), 75; Courtney Addison, this volume.

5 The Anthropocene Campus Melbourne (ACM18) built on other campus events in Berlin, Philadelphia, and elsewhere; it engaged participants in a range of lectures, field trips, and workshops in Melbourne and the wider area exploring "The Elemental." See "Anthropocene Campus Melbourne,"

Haus der Kulturen der Welt and the Max Planck Institute for the History of Science, https://archive.anthropocene-curriculum.org/pages/root /related-projects/anthropocene-campus-melbourne/.

6 For example, Cameron Allan McKean, "Telling Time through Lagoons of Human Waste at the Western Treatment Plant," *Anthropocene Curriculum*, 2018, https://www.anthropocene-curriculum.org/contribution/telling-time-through -lagoons-of-human-waste-at-the-western-treatment-plant; Jessica Cattelino, Georgina Drew, and Ruth A Morgan, "Water flourishing in the Anthropocene," *Cultural Studies Review* 25, no. 2 (2019); Alison Kenner, Aftab Mirzaei, and Christy Spackman, "Breathing in the Anthropocene: Thinking through scale with containment technologies," *Cultural Studies Review* 25, no. 2 (2019).

7 Timothy Neale, Thao Phan, and Courtney Addison, "An Anthropogenic Table of Elements: An Introduction," Society for Cultural Anthropology, https:// culanth.org/fieldsights/an-anthropogenic-table-of-elements-an-introduction.

8 See https://www.4sonline.org/a-cordial-invitation-to-the-table-of-elements/.

9 Donna Haraway, *Modest_Witness@Second_Millennium.FemaleMan©_Meets_ OncoMouse™: Feminism and Technoscience* (New York, NY: Routledge, 1997), 68; Joseph Dumit, "Writing the Implosion: Teaching the World One Thing at a Time," *Cultural Anthropology* 29, no. 2 (2014): 349.

10 See Thao Phan's chapter in this volume for further discussion of this event.

11 Gabrielle Hecht, "Interscalar Vehicles for an African Anthropocene: On Waste, Temporality, and Violence," *Cultural Anthropology* 33, no. 1 (2018): 135.

12 Alexis Rider, this volume.

13 Timothy Neale, this volume.

14 John Durham Peters, *The Marvelous Clouds: Toward a Philosophy of Elemental Media* (Chicago, IL: University of Chicago Press, 2015), 1.

15 Nicole Starosielski, "The Elements of Media Studies," *Media+ Environment* 1, no. 1 (2019): 4; Eli Elinoff, this volume.

16 For more, see Sasha Engelmann and Derek McCormack, "Elemental Aes- thetics: On Artistic Experiments with Solar Energy," *Annals of the American Association of Geographers* 108, no. 1 (2018); Derek McCormack, this volume.

17 Alexis Rider and Derek McCormack, this volume.

18 Cymene Howe, *Ecologics: Wind and Power in the Anthropocene* (Durham, NC: Duke University Press, 2019), 11.

19 Heather Davis and Zoe Todd, "On the Importance of a Date, or Decolonizing the Anthropocene," *ACME: An International E-Journal for Critical Geographies* 16, no. 4 (2017).

20 Zeynep Oguz, this volume.

21 Janelle Lamoreaux and Ayo Wahlberg, this volume.

22 Engelmann and McCormack, "Elemental Aesthetics."

23 Hannah Gibson and Sita Venkateswar, "Anthropological Engagements with the Anthropocene: A Critical Review," *Environment and Society* 6 (2015): 5–27.

24 Rider, this volume.
25 Engelmann and McCormack, "Elemental Aesthetics," 243.
26 Brad Bolman, this volume; Courtney Addison, this volume; Eli Elinoff, this volume; Scott Wark, this volume; J.R. Latham and Kate Seear, this volume.
27 Timothy Mitchell, *Carbon Democracy: Political Power in the Age of Oil* (Verso Books, 2011).
28 Hecht, "Interscalar Vehicles," 135.
29 Michelle Murphy, "Alterlife and Decolonial Chemical Relations," *Cultural Anthropology* 32, no. 4 (2017): 495; see also Addison, this volume; Tironi, this volume; and Desrochers-Turgeon, Saloojee, and Todd, this volume.
30 Murphy, "Alterlife," 496.
31 Michelle Murphy, "Some Keywords Toward Decolonial Methods: Studying Settler Colonial Histories and Environmental Violence from Tkaronto," *History and Theory* 59, no. 3 (2020): 376–84.
32 Flavio D'Abramo and Hannah Landecker, "Anthropocene in the Cell," *Technosphere Magazine* (2019). https://technosphere-magazine.hkw.de/p /Anthropocene-in-the-Cell.
33 Manuel Tironi, this volume.
34 D'Abramo and Landecker, "Anthropocene in the Cell."
35 Alexis Shotwell, *Against Purity: Living Ethically in Compromised Times* (Minneapolis, MN: University of Minnesota Press, 2016), 9.
36 Xan Chako and Neale, this volume.
37 Karen Barad, "After the End of the World: Entangled Nuclear Colonialisms, Matters of Force, and the Material Force of Justice," *Theory and Event* 22, no. 3 (2019).
38 Barad, *Meeting the Universe Halfway*, 73.
39 Hecht, "Interscalar Vehicles," 135.
40 Wark, this volume.
41 Engelmann and McCormack, "Elemental Aesthetics," 242.
42 Claire Colebrook, "We Have Always Been Post-Anthropocene: The Anthropocene Counterfactual," in *Anthropocene Feminism*, ed. Richard Grusen (Minneapolis, MN: University of Minnesota Press, 2017), 7.
43 Donna Haraway, *Staying with the Trouble: Making Kin in the Chthulucene* (Durham, NC: Duke University Press, 2016), 101.
44 Anna L. Tsing, et al., "Introduction: Haunted Landscapes of the Anthropocene," in *Arts of Living on a Damaged Planet: Ghosts and Monsters of the Anthropocene*, ed. Anna L. Tsing et al. (Minneapolis, MN: University of Minnesota Press, 2017); João Biehl and Peter Locke, "Ethnographic Sensorium," in *Unfinished: The Anthropology of Becoming*, ed. João Biehl and Peter Locke (Durham, NC: Duke University Press, 2017), 11.
45 Scerri, *The Periodic Table: Its Story and Its Significance*, 113.

46 Jeffrey Jerome Cohen and Lowell Duckert, "Eleven Principles of the Elements," in *Elemental Ecocriticism: Thinking with Earth, Air, Water, and Fire*, ed. Jeffrey Jerome Cohen and Lowell Duckert (Minneapolis, MN: University of Minnesota Press, 2015), 4.

47 Ann Laura Stoler, "Imperial Debris: Reflections on Ruins and Ruination," *Cultural Anthropology* 23, no. 2 (2008).

48 Alexander G Weheliye, *Habeas Viscus: Racializing Assemblages, Biopolitics, and Black Feminist Theories of the Human* (Duke University Press, 2014); Sylvia Wynter, "Unsettling the Coloniality of Being/Power/Truth/Freedom: Towards the Human, after Man, Its Overrepresentation – An Argument," *CR: The New Centennial Review* 3, no. 3 (2003).

49 See Stacy Alaimo, *Exposed: Environmental Politics and Pleasures in Posthuman Times* (Minneapolis, MN: University of Minnesota Press, 2016).

50 By attending to the "kerosphere," for example, Desrochers-Turgeon, Saloojee, and Todd (this volume) advocate for an anticolonial reading of Alberta's mining industries that might generate better relations with the land.

51 Chako, this volume.

52 Donna Haraway, *Modest_Witness@Second_Millennium.FemaleMan©_Meets_OncoMouse™*, 24.

53 For example, the lore of the Kulin Nation people in southern Australia, within what is commonly known as Victoria, describes the origins of fire in the rash actions of the crow Waa (also Wahn or Waang, Werpil).

54 See Cherkaev, Paxson, and Helmreich, this volume; Keck, this volume; Neale, this volume; Chacko, this volume.

55 Nicholas Mirzoeff, "Visualizing the Anthropocene," *Public Culture* 26, no. 2 (73) (2014): 213–32.

56 Phan, this volume.

1 1080

courtney addison

An animal skitters through leaf matter. The animal has four feet with small claws, and fur that you'd find to be soft if you touched it, except that you won't. You won't touch the animal, unless you were to gingerly take it by its tail and feel surprise at the dead weight of it. If you took the animal by the tail, once you'd thrown it away you would immediately and thoroughly wash your hands of it, the way advertisements and posters tell you to do (between each finger, lather the soap, dry properly), the way you never do, unless you've touched something truly polluting. The animal is polluting. It is a rat, and rats are pests. The rat's small body comes saddled with the spectre of disease, with the blind contagions of its ancestors, with the crimes of ancestors who colonized the country through which it is now travelling. It is led by its exceptional nose which can tell that there are small, warm, other animal bodies close by, soft-heaving with little breaths, feathered instead of furred. And because it is a rat and it is driven to eat, the animal will break into these nests or burrows that are so alive with little life, and it will crunch down on other bodies or through the shells of eggs, and as so many of its kind are doing this, all over the islands of Aotearoa, a whole army of bigger animals will mobilize and come forth with traps and baits to protect the feathered bodies and kill the furred.

The summer bridging 2018 and 2019 was a warm one for Aotearoa New Zealand. A marine heatwave pulled people towards the unusually warm coastal waters, while air temperatures sat 1.2°C above historical averages. It was the third warmest summer in the 110 years since record keeping began, and the country's flora responded to the heat.[1] In autumn, the beech and rimu trees that cover the South Island erupted in flowers, shedding as many as 15,000 seeds per square metre.[2] Scientists leaned out of helicopters and cut small branches from the trees in a kind of scientific divination. Reading these auguries, they concluded that it would be a mega-mast season, where trees, tussocks, and other foliage fall into step to synchronize a mass seeding event.[3]

Mast seasons set off a dangerous ecological chain of events in Aotearoa. Rodent numbers explode to take advantage of the abundant seed matter, and stoats move in to prey upon the rats and mice; when the seeds begin to rot, those predators turn to native birds to sustain their food intake. In 2013, local researchers developed a new model for predicting mast events, arguing that the best predictor of a mast was the mean temperature difference between two preceding years: a pronounced increase in temperature from one summer to the next is likely to trigger a mast event in the following year.[4] In anticipation of the mast, the Department of Conservation (DoC) undertook its most extensive pest control program to date, covering one million hectares, or 12 per cent of the country's total conservation land.[5] The main task of this operation involved the distribution by helicopter of green bait pellets laced with sodium fluoroacetate, a poison more commonly known by its original serial number: 1080.

Here, I take 1080 as a site for exploring the politics of the Anthropocene in Aotearoa. Like other chemical agents, it is a useful "linking figure"[6] whose loops we can follow through military industry, colonization, multispecies bodies, waterways, protests, laboratory assays, and legislation. But, as Murphy[7] reminds us, "effects and injuries are not chained" neatly from production to consumption. Indeed, the injuries of 1080 are an aftermath themselves, following a much longer and more injurious history of colonization that killed, hurt, and dispossessed both human and nonhuman populations in Aotearoa. This substance patterns life and death across space and species, and has become the heart of one of this country's most volatile political controversies. In what follows, I trace these patterns and links, the contemporary and historic discontents of 1080, and how this poison holds both curative and calamitous possibilities in play. Like the Greek *pharmakon* that represented both remedy and poison, 1080, and our attempts to write it, are simultaneously imitative and generative, deadly and life-giving.[8] Observing how the poison's value and danger are negotiated in Aotearoa highlights more broadly the necropolitical underpinnings of the country's ongoingly settler-colonial political order.

The story of Aotearoa goes back over 500 million years, to when ash and sediment bonded at the fringe of the Gondwana supercontinent. 85 million years ago, a swell of hot subterranean rock saw this fringe break away. The new continent of Zealandia drifted out and down into Te-Moana-Nui-a-Kiwa, the Pacific Ocean, where it settled, mostly underwater, gently atop the seam of the Pacific and Australasian tectonic plates. Recently, over the last 1.8 million years or so, the slow grate of these plates has pushed bits of Zealandia up out of the water. Mountains have risen, rock has become country, and glaciers have choreographed a new terrain.

Over time, birds flew to the country and made it their home, and other small creatures found their way on currents of wind or ocean. Through adaptive radiation they evolved to fill the country's many ecological pockets, while flora grew towards the insects available. There were no naturally pollinating vertebrates in Aotearoa and the only terrestrial mammals were a few species of native bat.[9] Polynesian navigators arrived by waka, and had made their home in Aotearoa by 1300AD. They bought kumara and taro, as well as the kiore (rat) and kuri (dog). During this period, some pre-existing species were killed for their meat and feathers, in some cases to the point of extinction, while many others were treasured as pets, ancestors, status symbols, and allies.[10]

When the British colonized Aotearoa they brought with them myriad foreign species for food, sport, and fur. Possums were introduced to establish a fur trade, trout and pheasants for hunting, and rabbits for food. The "grasslands revolution" marked mass deforestation as land dressed in native bush was converted to farm cattle and sheep.[11] The land where colonizers arrived was thus not innocent of human impact, but it underwent an exponential intensification of land-clearing and species loss as a result of British settler-colonization. In a country that had previously been devoid of all predators, the influx of egg- and mammal-eating creatures hit native species hard. Birds like the kākāpo and kiwi had long ago lost the ability to fly, adapting instead to live on the ground and in burrows; they became easy prey for stoats and dogs. Early reports depict the huge kākāpo parrots as sitting targets.[12] They did not know to move when a new creature approached them; nothing had ever hurt them before.

Of course, the acceleration of death that accompanied Aotearoa's colonization was not confined to birds and flora. The Māori population had reached a nadir by 1900, following a century of resistance to British military aggression and land confiscation.[13] Although the population increased during the twentieth century, official government rhetoric made "the outrageous claim that New Zealand was the one place on the planet where racial harmony prevailed," discursively erasing the country's many ongoing social inequities.[14] This, then, is the Aotearoa that 1080 entered into upon its adoption in the 1950s. The landscape had been refashioned according to British ideals in a bid to assert colonial ways of life and increase agricultural productivity.[15] Farmers around the country were enjoying the "long boom."[16] Agricultural productivity had been fuelled by superphosphates mined from Banaba and Nauru,[17] refrigeration technology facilitated the export of meat and dairy as well as wool,[18] and Commonwealth trading rights were protected via the Ottawa agreement.

Farmland, however, was being ruined by rabbits, which burrowed destabilizing underground networks and outbred all attempts at control. Legislative amendments in the 1940s saw the government disincentivize rabbit hunting, which had proven ineffectual at controlling their populations, in favour of what is widely termed a "killing policy" directed by the Rabbit Destruction Council. It was in this context that 1080 was brought to Aotearoa, where it joined a toxic arsenal of pest control measures, including "the distribution of jam poisoned with phosphorus, the fumigation of burrows with chloropicrin and the dropping of poisons from planes."[19]

The feel of warfare sits heavily over this coordinated killing policy, and prefaces the language of war[20] and securitization[21] that continues to define conservation in Aotearoa. Indeed, the distribution of 1080 by helicopters followed the trial and adoption of aerial fertilization using air force planes and personnel retired from the Second World War.[22] Over the past seventy years the specific species 1080 is used on and, indeed, the broader rationale for the poison, has shifted: whereas it was initially used on rabbits for their ruination of farmland, today 1080 is primarily used to kill the rats and mustelids that threaten the country's many vulnerable native birds. However, the poison's continuous presence indicates its centrality to the necropolitics of this place, in which the calculated cultivation of native non-human life depends upon the calculated exploitation and destruction of select non-native species. As Dutkiewicz[23] argues, not all introduced species are equally unwelcome: the economically valuable species that the country's primary production sector relies upon never attract the necropolitical force of those deemed pests, despite conforming to a similar criterion of non-nativeness. Furthermore, the otherness of pest species and their predation upon native species is metabolized into a kind of symbolic value that legitimates "killing [a]s the normal form of engagement"[24] and naturalizes settler colonial relations to the land. In this context, where the lives of Aotearoa's human population and native species depend upon selectively propagated death, 1080 becomes a *pharmakon*: literal poison and putative remedy, agent of life through death, morally ambivalent but ostensibly indispensable. Where elsewhere, the toxicities that characterize the Anthropocene emerge as externalities and side-effects of industry and extractivism, in Aotearoa toxicity is the point.

As a commercial product, 1080 arose from the specific geopolitics of America's role in the Second World War. Preparing to deploy to the Asia-Pacific region, US medics anticipated encountering endemic tropical diseases, and military personnel soon began to suffer from scrub typhus, or tsutsugamushi disease.[25] At the same time, military developments in Europe began to choke America's supply of imported rodenticide, prompting the US government to instigate a number of "crash projects" to develop local alternatives. In "an illustration of the advantages of wartime cooperation,"[26] the Office of Scientific Research and Development funded the US Fish and Wildlife Service (FWS) in a search for these new compounds, supported by various public health and military agencies, including the British Commonwealth Scientific Office and the US Army's Chemical Warfare Service. It was the entwining of war and wildlife that conceived 1080.

The FWS began testing 1080 on albino lab rats in mid-1944, and quickly expanded testing to other rodents and prairie dogs before moving into field tests. Two months into the process, the substance received a classification of Secret under the Espionage Act, although in a matter of months this was downgraded to Restricted, and then Open.[27] In 1945 a short report in *Science* announced the discovery of "a promising new rodenticide."[28] The poison's promise stemmed from its high and quickly achieved toxicity to rodents and palatable taste compared to other poisons. It was chemically stable and did not appear to irritate the skin, improving its ease of use. This is precisely the balance that its developers were looking for: lethal efficacy against rodent bodies, and tolerability to human ones. However, the toxicity that made it such an effective rodenticide was also one of the product's risks: in addition to rodents it was fatal upon ingestion to cats, dogs, and people. Moreover, the substance was so fast-acting that it would be difficult to save anything that accidentally consumed it. The government agencies that developed 1080 resolved to buy the entire supply of the product (then being produced by Monsanto), until regulation could be put in place to limit its availability to lay persons.[29] Questions of control and responsibility adhered to 1080 from the earliest stages of the poison's life.

Today, Aotearoa New Zealand is the world's largest consumer of 1080, purchasing 80 per cent of the total global supply.[30] The poison is manufactured by Tull Chemical Company, in Alabama, and imported by a small number of authorized buyers. When it reaches local shores, the raw sodium fluoroacetate is processed into baits that are then used in pest control operations throughout the country. Activists harness this offshore provenance to argue that 1080 does not belong in Aotearoa, that it does violence to the "100 per cent pure," "clean green," "New Zealand natural" identity of the nation. These slogans have a double significance. They index the literal branding of Aotearoa by its national tourism agency and thus the economic value that nature generates. But they also suggest ideals of an untouched, pre-colonial Aotearoa, envisioned as a paradise of birds.[31] Like conservation more widely, these affect-laden appeals to nature demand critical appraisal; just as conservation has been critiqued for naturalizing settler colonial occupation of Indigenous lands,[32] "imperial nostalgia" for the lifeforms transformed through colonialism conceals the complicity of colonial destruction, recoding it instead as innocent.[33] These same sentiments are advanced both in favour and against 1080.

Indeed, the poison is highly controversial in Aotearoa and provokes an intensity of political and social opposition that is unusual for this country. In 2019, the Department of Conservation (DoC) committed over NZ$10 million to staff security over a four-year period, in response to death threats against conservation workers. Five years earlier, envelopes of infant formula laced with 1080 were mailed to Fonterra (a New Zealand company that exports 30 per cent of the world's dairy) and Federated Farmers, along with threats that the anonymous sender would poison export supplies. This stunt, which was eventually traced to a businessman who sold a rival pest control product, capitalized on perceived oppositions between the purity of Aotearoan biomatter and the toxic potency of 1080.

Most of the New Zealand population will never encounter 1080 in person. As a result, opponents of the poison have to work to make its effects more widely legible. Two aesthetic tactics in particular seem to be popular among protestors: visual displays of death and aural productions of silence. An extensive and unreliable set of images circulate online testifying to the effects of 1080. One widely shared photograph portrays dozens of dead kiwi, their feet tied and beaks askew. The birds pictured had been killed by dogs and cars over a three-year period, and the image had been shared by a small town conservation group that was monitoring kiwi and trapping predators in the Tongariro National Park. Once online though, the photograph was rapidly repurposed by activists, who attributed the deaths to the poison and supported their claims with decontextualized fragments of quotes from DoC staff. Other images similarly shock with the sheer number of dead bodies or their grotesque expressions of death. Foaming mouths. Bent necks. Bloodied hides. All seek to evoke the supposed violence unleashed by the poison. DoC, in comparison, have made a photo of a stoat with a dead baby bird in its mouth as the landing image of the webpage dedicated to debunking 1080 myths and hoaxes. The fledgling's oversized beak and hopelessly limp form contrast with the menacing eye contact made by the stoat as it bounds directly towards the photographer.

Those who oppose the poison also describe walking through the bush after a "drop" and hearing only eerie silence. Videos and audio-recordings uploaded to anti-1080 Facebook pages claim to capture the silence of the bush after a 1080 operation, and others, the chorus of birds in areas that have gone untreated. Proponents argue exactly the opposite, describing how the bush seems to come back to life in the months after pest control operations. The validity of activists' aural renderings is deeply suspect: there is no way to tell where or when the recordings were made, and if there *is* raucous birdsong in untreated areas, then any causal relationship likely runs the other way (i.e., areas where birds are thriving do not warrant 1080 use, or are accessible enough that trapping and other pest control methods suffice). Nonetheless, placed beside each other the vivid depictions of silent and populous bush conjure a kind of spectrality that powerfully alludes to extinction. The motif of silence recurs in the submissions made to the Environmental Protection Authority during their 2008 review of 1080 use, with both detractors and advocates of the poison either invoking or disputing its silencing effects.[34] McCorristine and Adams argue that ecological relations are by definition so dense that a single species' extinction can summon a whole chorus of hauntings.[35] The example these UK-based scholars give is the New Zealand lancewood, a tree that retains its adaptations to the long-extinct megafaunal bird, the Moa. Such "ecological hauntings"[36] seem to proliferate loss, as the disappearance of one animal unfolds exponential losses of relations. Even if the silences anti-1080 activists share are believed, though, they continue to divide. Is the appropriate response to a silent forest less poison, or more?

There is a common impulse shared by anti-1080 activists when they conjure a silent, haunted bush, and conservationists when they recall the paradise of birds that Aotearoa once resembled. Even around this most contentious agent of species preservation, debates over conservation in Aotearoa embody both a restorative[37] and imperialist[38] nostalgia – one that invokes supposedly better times without fully accounting for how we got from there to here. Historian Svetlana Boym's restorative nostalgia regards the past "not [as] a duration but a perfect snapshot," imposing a singularity and stasis that is necessary for restitution.[39] The interaction of these two forms of nostalgia generates the "preservationist ethic" that has come to define conservation in Aotearoa.[40] The compound 1080 is central to this preservationist impulse, appearing to embody the very logics of salvation through war from which it initially sprung.[41]

The ancient Greek *pharmakon* traditionally denoted a substance that held in play the possibility of both remedy and poison, but was adopted by Plato in his lamentation of the written word. For Plato, writing seemed to be an aid to memory, but was also always complicit in memory's erosion, substituting live remembering with inert "monuments," resigned to imitation. Philosopher Jacques Derrida resuscitated the term in his essay "Plato's Pharmacy," suggesting that it is the suspense between the life-giving and deathly, it's always-perhaps-both-at-once, that lends the *pharmakon* concept its "resources of ambiguity."[42] If ever there were a time for such resources, the Anthropocene is it; at a time where the very technologies and substances that were supposed to give, enrich, or extend life turn out to be relentlessly compromised, the *pharmakon* helps us account for the many symbolic and material paradoxes of the present.

Like Plato's *pharmakon*, 1080 collapses life and death into each other. It is precisely the deathliness of 1080 that allows it to be remedial, and the remedy is always of and flushed with death. And so here I have sought to draw out the ambiguities of 1080, by mapping "the movement, the locus, and the play"[43] of this poison through space and time, but also via the act of naming it a *pharmakon*, suggesting that we hold in view its many contradictions. Indeed, many Aotearoa-based conservationists do precisely this. In my interviews with conservation scientists, many have spoken about the poison as a necessary evil, or something they didn't like but knew we had to use. The compound 1080 may promise to save our native species, but it can only do so by killing selectively among those that were introduced. This apparently unavoidable compromise, inextricable from its cloak of contrition, bespeaks a deeper necropolitical logic that resonates more widely with the predicament of the Anthropocene: How do we organize our killing, and square it with those pasts that we will acknowledge, and those we will not?[44]

Derrida highlights an additional conceptual valence, where the *pharmakon* bleeds into the *pharmakos*. In ancient Greece, the *pharmakos* was a human scapegoat who was ritually expelled from the city and put to death in an act of purification intended to safeguard the polis. As the representative of the outside world the *pharmakos'* life within the city represented a mortal threat, and their elaborate killing suggested the promise of a cure. The pest fills a similar role in Aotearoa. Within the necropolitical workings of power, it is not just that the pest is killable, but that its killing can be put in service of securing the state. Furthermore, drilling into this language highlights the second function of the scapegoat: to take the fall for another, to bear the burden of blame for another's wrongdoings. Talking about how pest species should be killed is a way of talking around colonial histories, one enactment of the "relentless cultural disposition" towards moving forward and progress that geographer Lesley Head describes.[45] And as Davis and Todd remind us, we cannot hope to rise to the challenges of the Anthropocene if we cannot acknowledge its oppressive roots, nor how these reach differently into the dirt of diverse global locales.[46]

To write about 1080 is to combust two forms of *pharmakon*, the poison and the language from which it is constantly escaping. Writing is always a task of fixing something alive to the page, and writing about 1080 – writing this essay – immobilizes something that is fundamentally volatile. In this way, the writing of 1080 fits with the same epistemic tradition as charting the Periodic Table: an effort to grasp and define something ever-changing through the medium of sticky marks that we know as language. When Mendeleev created his Periodic Table of Chemical Elements, he kept spaces for those elements that he anticipated but that were not yet known; the regularity of elemental properties allowed this calculated anticipation, and created space for it to be etched into the page. These blank spaces in Mendeleev's table are arguably its most powerful feature. If the neatly charted elements purport to speak for all of space and time, the blanks remind us that what mattered in 1869 is not necessarily what matters now.

In 1869, sodium fluoroacetate was quietly leaching from plants across Brazil and Australia; today it is sown with utmost precision by helicopter across a different landmass entirely. In 1869, the New Zealand Wars were approaching their final years, at the end of a decade marked by the mass confiscation of Māori land by Pākehā settlers. This history is often caught in a long shadow of Pākehā memory, which continues to buff out the complexities and violence of Aotearoa's colonization.[47] In such contexts, it takes more than the language of technoscience to "grapple with the expansive chemical relations of settler colonialism."[48] Following 1080 attunes us to both the distinct temporality of the Anthropocene in this place, with its two historic waves of human arrival, and the particular alliances of environment, industry, and human-nonhuman life that precipitated the species loss and land change which arguably embody the Aotearoa Anthropocene.

Ultimately, to think of 1080 as *pharmakon* is to raise the question of how we regard the past. Like Plato's writing, like Boym's snapshot, 1080 is fundamentally imitative: it seeks to restore, but is resigned to restoring "only a kind of ghost" of pre-colonial nature.[49] Read as "a state of impairment,"[50] the Anthropocene might appear to require new remedies, but following the histories of 1080 and Aotearoa reminds us of the deep ambivalence of our current medicine chest. In a country with "a paradoxically rich culture of extinction,"[51] where ecological hauntings teeter perpetually on the brink of metastasis, 1080 is at once a scapegoat, a compromised cure, and an agent of death.

NOTES

1 NIWA, "New Zealand's 3rd-Warmest Summer on Record," 2019, https://niwa.co.nz/sites/niwa.co.nz/files/Climate_Summary_Summer_2018–19-NIWA.pdf.

2 DOC, "Monitoring Forest Seeding," 2019, https://www.doc.govt.nz/our-work/tiakina-nga-manu/forest-seed-monitoring.

3 DOC, "Mega Mast Confirmed for New Zealand Forests," 2019, https://www.doc.govt.nz/news/media-releases/2019/mega-mast-confirmed-for-new-zealand-forests/.

4 Dave Kelly et al., "Of Mast and Mean: Differential-Temperature Cue Makes Mast Seeding Insensitive to Climate Change," *Ecology Letters* 16, no. 1 (2013): 90–8.

5 DOC, "Mega Mast Confirmed for New Zealand Forests."

6 Nicholas Shapiro and Eben Kirksey, "Chemo-Ethnography: An Introduction," *Cultural Anthropology* 32, no. 4 (2017): 481–93.

7 Michelle Murphy, "Chemical Regimes of Living, " *Environmental History* 13, no. 4 (2008): 695–703.

8 Jacques Derrida, *Dissemination*, trans. Barbara Jo (London: The Athlone Press, 1981).

9 Matt McGlone, "Evolution of Plants and Animals," Te Ara – the Encyclopedia of New Zealand, 2007, http://www.teara.govt.nz/en/evolution-of-plants-and-animals.

10 Annie Potts, Philip Armstrong, and Deidre Brown, *A New Zealand Book of Beasts: Animals in Our Culture, History and Everyday Life* (Auckland: Auckland University Press, 2013).

11 Tom Brooking and Eric Pawson, "Silences of Grass: Retrieving the Role of Pasture Plants in the Development of New Zealand and the British Empire," *Journal of Imperial and Commonwealth History* 35, no. 3 (2007): 417–35.

12 Charlie Douglas, *Mr. Explorer Douglas: John Pascoe's New Zealand Classic*, revised by Graham Langton (Christchurch: Canterbury University Press, 2000).

13 James Belich, *The New Zealand Wars and the Victorian Interpretation of Racial Conflict* (Auckland: Auckland University Press, 1986).

14 Peter Adds, "New Zealand's Treaty of Waitangi Reconciliation Processes: A Māori Treaty Educator's Perspective," in *Reconciliation, Representation and Indigeneity: "Biculturalism" in Aotearoa New Zealand*, eds. Peter Adds, Brigitte Bönisch-Brednich, Richard S. Hill, and Graeme Whimp (Universitätsverlag Winter, 2016), 19–25.

15 Brooking and Pawson, "Silences of Grass."

16 Julia Haggerty, Hugh Campbell, and Carolyn Morris, "Keeping the Stress Off the Sheep? Agricultural Intensification, Neoliberalism, and 'Good' Farming in New Zealand," *Geoforum* 40, no. 5 (2009): 767–77.

17 Katerina Teaiwa, *Consuming Ocean Island: Stories of People and Phosphate from Banaba* (Bloomington: Indiana University Press, 2015).

18 Rebecca J.H. Woods, "Breed, Culture and Economy: The New Zealand Frozen Meat Trade, 1880–1914," *The Agricultural History Review* 60 no. 2 (2012): 288–308.

19 Rachael Egerton, "Unconquerable Enemy or Bountiful Resource? A New Perspective on the Rabbit in Central Otago," Environment and Nature New Zealand, 9, no. 1 (2014): n.p.

20 Courtney Addison, "Compound 1080 (Sodium Monofluoroacetate)," *Theorizing the Contemporary, Fieldsights.* Society for Cultural Anthropology (website), eds. Timothy Neale, Thao Phan, and Courtney Addison. From the series An Anthropogenic Table of Elements, 27 June 2019), https://culanth.org/fieldsights/compound-1080-sodium-monofluoroacetate.

21 Thomas Isern, in James Beattie, "Biological Invasion and Narratives of Environmental History in New Zealand, 1800–2000," *Invasive and Introduced Plants and Animals: Human Perceptions, Attitudes and Approaches to Management,* eds. Ian D. Rotherham and Robert A. Lambert (London, Washington: Routledge, 2012).

22 Teaiwa, *Consuming Ocean Island.*

23 Jan Dutkiewicz, "Important Cows and Possum Pests: New Zealand's Biodiversity Strategy and (Bio)Political Taxonomies of Introduced Species," *Society and Animals* 23, no. 4 (2015): 379–99.

24 Dutkiewicz, "Important Cows," 385.

25 Alison Luce-Fedrow et al., "A Review of Scrub Typhus (Orientia Tsutsugamushi and Related Organisms): Then, Now, and Tomorrow," *Tropical Medicine and Infectious Disease* 3, no. 1 (2018): 8; Justus C. Ward, "Rodent Control with 1080, ANTU and Other War-Developed Toxic Agents," *American Journal of Public Health* 36 (1946): 1427–31.

26 Ward, "Rodent Control with 1080": 1428.

27 Guy Connolly, "Development and Use of Compound 1080 in Coyote Control, 1944–1972," *Proceeding of the 21st Pest Conference* 21, no. 12 (2004): 221–39.

28 Edwin R. Kalmbach, "'Ten-Eighty,' a War-Produced Rodenticide," *Science* 102, no. 2644 (1945): 232–3.

29 Connolly, "Development and Use of Compound 1080."

30 Landcare Research, "Section Five: International Considerations of 1080," in *1080 Reassessment Application* (Accessed 12 February 2020, 2006), 446–57.

31 Cameron Boyle, "Remembering the Huia: Extinction and Nostalgia in a Bird World," *Animal Studies Journal* 8, no. 1 (2019): 66.

32 Jonathan Goldberg Hiller and Noenoe K. Silva, "Sharks and Pigs: Animating Hawaiian Sovereignty against the Anthropological Machine," *South Atlantic Quarterly* 110, no. 2 (2011): 429–46.

33 Renato Rosaldo, "Imperialist Nostalgia," *Representations* 26 (1989): 107–22.

34 Enviromental Protection Authority, "Appendix T: Summary of Submissions," 2008.

35 Shane McCorristine and William M. Adams, "Ghost Species: Spectral Geographies of Biodiversity Conservation," *Cultural Geographies* 27, no. 1 (2020).

36 McCorristine and Adams, "Ghost Species," 106.

37 Svetlana Boym, *The Future of Nostalgia* (New York: Basic Books, 2001).

38 Rosaldo, "Imperialist Nostalgia."

39 Boym, *The Future of Nostalgia*, 49.

40 Franklin Ginn, "Extension, Subversion, Containment: Eco-Nationalism and (Post) Colonial Nature in Aotearoa New Zealand," *Transactions of the Institute of British Geographers* 33, no. 3 (2018): 335–53.

41 The notion of war as a means to salvation is, of course, articulated by Achille Mbembe in his treatise on necropolitics. He writes: "The sacramentalization of war and race in the blast furnace of colonialism made it at once modernity's antidote and poison, its twofold *pharmakon*" (6).

42 Derrida, *Dissemination*, 97.

43 Derrida, *Dissemination*, 127.

44 Thom Van Dooren, "A Day with Crows: Rarity, Nativity and the Violent-Care of Conservation," *Animal Studies Journal* 4, no. 2 (2015): 1–28.

45 Lesley Head, *Hope and Grief in the Anthropocene* (London, New York: Routledge, 2018).

46 Heather Davis and Zoe Todd, "On the Importance of a Date, or, Decolonizing the Anthropocene," *ACME, An International Journal for Critical Geographies* Vol. 16, no. 4 (2017): 761–80.

47 Vincent O'Malley and Joanna Kidman, "Settler Colonial History, Commemoration and White Backlash: Remembering the New Zealand Wars," *Settler Colonial Studies* 8, no. 3 (2018): 298–313.

48 Michelle Murphy, "Alterlife and Decolonial Chemical Relations," Society for Cultural Anthropology 32, no. 4 (2017): 497, https://doi.org/10.14506 /ca32.4.02.

49 Derrida, *Dissemination*, 148.

50 Cymene Howe, *Ecologics: Wind and Power in the Anthropocene* (Durham: Duke University Press, 2019).

51 Boyle, "Remembering the Huia," 68.

2 CARBON

timothy neale

It is a short walk from my rental car to the conference centre at Charles Darwin University, but by the time I arrive at the registration desk for the 2020 North Australian Savanna Fire Forum my shirt is soaked with sweat. Locals have been talking about what a disappointing Wet season it has been here in Darwin and across the Northern Territory. There have been few storms to date, but as my shirt attests there is no escaping the elements; the familiar monsoonal humidity and sweltering heat still set upon you the moment you leave air conditioning. The people milling around registration are a mix distinctive to the field of savanna fire management – scientists, rangers, helicopter contractors, lawyers, gas industry lobbyists, carbon traders, and federal regulators – and in the churn of salutations and exchanged business cards, the tolling bell intended to signal the start of the day's formal proceedings is ignored for almost half an hour. Over the three years that I've attended the Forum it has consistently grown, and this year there is hardly a spare seat in the windowless 400-seat auditorium as a representative of the local Larrakia Indigenous custodians, or Traditional Owners, welcomes us. The first presentation, by a nervous man from the national Bureau of Meteorology, relays in graphs and maps what most present already know too well (see figure 2.1). Last dry season, from May to October, fire danger indexes were consistently registering values in the highest two deciles, and at times the highest readings on record. Three months into this wet season, the rainfall has been consistently the lowest readings on record in many regions. An older fire manager leans over and summarizes it for me with characteristic concision: "So, it's been pretty dry and windy."

While movements and distributions of wind, rain, and heat are concerns for anyone engaged in the management of flammable savannas, whether

Figure 2.1. Attendees at the 2020 North Australian Savanna Fire Forum (credit: Aneeta Bhole/SBS News)

in north Australia or elsewhere, the elemental entanglements that are the focus of the Forum are less meteorological or climatic than chemical. As is apparent from the event's schedule, participants, and sponsors, "carbon" is the keyword around which fire management is increasingly organized economically and socially here. Approximately 15 per cent of the tropical savanna's 2 million square kilometres are now managed for "carbon farming" or carbon emissions abatement projects. What this involves, among other practices, is lighting fires in the early dry season, when they typically burn less intensely, in order to reduce fires in the late dry season, when they produce more emissions and consume more organic matter. Using satellite data on the timing and extent of fires, and models of their biophysical effects, projects are issued with credits for the calculated emissions abated in a given year. Put simply, one's carbon crop is the surplus between the emissions from the most recent year and the yearly average of emissions from a past baseline period, before management. The vast majority of the credits "farmed" in this way are purchased by the Australian government on behalf of a nation leading the world in carbon emissions per capita, and by carbon-intensive companies such as airlines and oil and gas multinationals. Since its founding in 2012, this combustive economy has grown to AUD\$20 million in income every year, with roughly 75 per cent of all credits coming from projects on Indigenous peoples' lands. Nonetheless, savanna fires in total contribute less than 5 per cent of the nation's reportable annual emissions.[1]

As in previous years, I spent the days before the Forum with a group of Darwin-based bushfire scientists, catching up on gossip and hearing about current conflicts in their blended personal and professional worlds. Repeated exposure has made me literate in some of the more arcane battles – including ongoing disputations about the validity and workability of a new method to calculate the carbon sequestered in assorted forms of woody biomass – many of which unfold in the pages of scientific journals. For these reasons, I am surprised by the Forum's intense characterizing carbon credits generated by Indigenous companies on Indigenous lands as a "premium carbon product" or "premium quality offsets." This emphasis is first demonstrated by the launch of a promotional video made by the Australian government and Qantas Airways. The lights dim and we watch footage of majestic outback landscapes – red dirt and dramatic escarpments – interspersed with comments from different Aboriginal rangers and the famous Aboriginal actor Ernie Dingo. The "good news," Dingo declares at the video's end, is that these projects are "helping to reduce emissions, benefit community and businesses, and help the environment."[2] Later, after lunch, two non-Indigenous executives of Indigenous organizations present a keynote lecture styled on the television show *Mythbusters*. Their intent, as the ten myths they target make plain, is to dispel "rumours" that such projects harm biodiversity, are not "traditional" or "Indigenous," or incentivize proponents to burn too much and too frequently. Having avowedly "busted" eight of their ten myths they conclude that "it is important to get the story out there, but make sure it is the right story!"

Such attempts to protect the premium value of carbon credits from Indigenous fire management projects present interesting departures and consistencies with how carbon – the new "metric of the human"[3] – has been theorized to date. In one sense, these narrative endeavours accord with the many excellent critical accounts scholars have given of carbon markets, characterizing how they "make things the same" across material, spatial, and temporal differences.[4] Like the other iterations that have emerged in the aftermath of the 1997 Kyoto Protocol, the Australian carbon market relies on turning diverse forms of activity in disparate locations into something commensurable or, more simply, a commodity – a carbon credit, like those discussed at the Forum, which can be counted and traded because of agreed rules of standardization and property rights. The centrality of commodification to the whole endeavour is signalled in the ubiquity of the element carbon, which functions, like the dollar standard in global finance, as the base currency through which all others are converted. For example, carbon credits from savanna fire management are certified on the basis of the emissions – specifically methane (CH_4) and nitrous oxide (NO) – they abate; these emissions have a known equivalence or "exchange rate" with carbon dioxide in terms of their radiative effect on the Earth's climate. The creation of global and national carbon accounts has required nations to "reterritorialise" the climate as a zone of policy and action, and, complimentarily, has also required the establishment of multiple methods and infrastructures to ensure carboniferous credits and debts are measured, formatted, and exchanged with consistency between contexts like any other piece of property.[5]

The formations created to ensure carbon emitted, sequestered, stored, or abated can be staged in terms of "stable and singular" data are heterogeneous, widely diffused and formidably technical.[6] Mobilizing carbon as a commodity domestically or internationally requires, to give a brief itinerary, methods for counting baseline carbon "pools" (e.g., the vegetation in savanna landscapes), methods for measuring the effects of natural and anthropogenic processes on these pools (e.g., burning them at different times of year), methods for counting these effects (e.g., reading infrared and photographic imagery from satellites), and methods for auditing these methods to ensure their consistency and reliability. At the same time, these interlinked methods have to be made consistent with others participating in the same market, ostensibly producing the same commodity, whether from converting cropland to pasture, planting native trees and shrubs, reducing nitrous oxide emissions from irrigated cotton, reducing the methane emitted from cattle, avoiding land clearing, or some other recognized and regulated means. This is not to mention the legal arrangements, labour forces, government policies and agencies, private equity, land holders, and private traders also entangled in efforts to make commodities such as Australian Carbon Credit Units (or ACCUs). As the sociologist Donald MacKenzie has argued, carbon markets are conditioned not only by the politics of nation-states, but also by the esoteric "subpolitics" of their manifold technicalities.[7]

For many political economy scholars, though, what is important about carbon markets is that they all "make things the same" to *keep things the same* politically and economically. While the implementation of cap-and-trade, tax based and voluntary carbon schemes are often framed by politicians and news media as revolutionary interventions – whether for the better or worse – they are seen by many scholars as principally functioning to maintain hegemony. Carbon credits are, from this perspective, affordable licences to pollute, devised by industries and governments reliant on unsustainable resource extraction and purchased to protect their public image and excuse themselves from any obligation to fundamentally transform.[8] From this perspective, to quote Heidi Bachram, such schemes "build the illusion of taking action on climate change while reinforcing current unequal power structures" and their underlying racialized logics.[9] This is "carbon colonialism," in Bachram's terms, as various schemes force or coerce people in poorer nations or regions to transform their lives, and sometimes lose access to livelihoods and key resources, in order to offset the emissions-intensive lifestyles of people in wealthier nations and enrich the intermediaries in between. At the same time, this can be seen not simply as the securing of existing power structures but also their extension, because carbon markets seek to integrate ecologies, activities and actors outside existing capitalist economies.[10] What needs to be foregrounded, as the sociologist and activist Larry Lohmann has repeatedly argued, is that carbon markets *have not* significantly reduced the primary driver of climate change: the extraction and combustion of fossil fuels.[11]

But while we can look to carbon schemes and focus on their commodifying character, or their securing of established political economies, an exclusive focus on these dimensions obscures their immense social productivity and dynamism. Commensuration creates new relations among disparate and remote things, to the point that one might now easily discover the total greenhouse gas emissions released in the production of a can of beer – its "carbon footprint" – or compare these emissions with those released in the production of a totally different commodity, like a gigabyte of cloud data storage. This kind of commodity-chain analysis, in which the climatic burdens and benefits of different commodities are revealed, weighed, and compared, has become fully routinized in many contexts. But, as signalled at the Forum, carbon markets are also socially and materially generative, creating new ecologies, new institutions, new infrastructures, and new types of social relations, ostensibly in aid of remaking humans' relations with the climate.[12] At a conceptual level, as Jerome Whitington notes, the Kyoto Protocol founded "an imaginative space of global atmospheric relations," a discursive space capable of connecting disparate entities and actors in terms of their mutual Earthly entanglement in carbon.[13] And as ethnographies of various carbon crediting projects in China, Congo, Mexico, Uganda and elsewhere demonstrate, making things commensurate requires constant improvising with contingencies, uncertainties and imperfections across multiple scales.[14] Commensuration is hard work.

Carbon credits would therefore seem like an ideal type of "interscalar vehicle," in historian Gabrielle Hecht's terms.[15] As Hecht argues, current debates in the hard sciences regarding the temporal origins and geological signature of the Anthropocene epoch should be seen as an invitation to also construct less universalizing accounts. This would be a way of "putting the Anthropocene in place" while keeping disparate human and nonhuman entities in the same conceptual framework, telling situated stories of the materialities that constitute the current ecological crisis "so as to better grasp the kinds of entanglements – and futures – people face in our treacherous times."[16] Curiously, the promoters of carbon markets appear to invite exactly this mode of scalar storytelling. When I go to my local café in Melbourne, for example, branding and promotional materials encourage me to see my individual consumption of a reusable coffee cup as linked to environmental interventions at a range of scales, including local recycling economies, renewable energy projects in developing nations, and thereby global climate change. When I board a Qantas domestic flight to Darwin, the in-flight magazine advises me that by choosing to "Fly Carbon Neutral ... your money goes directly to carbon offset projects" like that on Dambimangari peoples' Country in the northern Kimberley, pictured in a glossy double-page spread. My personal emissions are simultaneously, if only infinitesimally, profiting specific Aboriginal individuals and mitigating planetary climate change. The crafting of such stories suggests that any attempt to critically narrate carbon's interscalar travels has to contend with a crowd of storytellers.

This brings us back to the 2020 Forum and the value of carbon credits from Indigenous fire management projects. The existence of such premium offsets suggests that these elemental markets, in which carbon is exchanged and global networks are structured, are not produced solely through narratives of commensurability. Such credits are "boutique," "more than a mere carbon reduction" or, as one trader put it to me, "not a strict commodity."[17] They contain something extra and they cost something extra. To indulge in some brief and approximate quantification: at present, the Australian carbon market is dominated by the federal government's AU\$2.25 billion Emissions Reduction Fund (ERF), which holds reverse auctions to purchase credits at the lowest possible cost (historically, AU\$10–AU\$14 per tonne of carbon abated). A voluntary market also exists, representing roughly 4 per cent of the total market, and while Indigenous fire management projects represent 4 per cent of credits contracted through the Emissions Reduction Fund, they represent approximately 20 per cent of the credits in the voluntary market. Further, credits from these projects regularly find prices in the voluntary market two to three times higher than through the ERF and feature prominently in the marketing materials of their purchasers. Thus, while the credits generated from Dambimangari peoples' Country represent under 5 per cent of Qantas' carbon offsets, their premium quality is realized in a high market price and the high marketing value of appearing to simultaneously support Indigenous peoples, biodiversity and climate change action. The complicating factor is that these other effects, often described as "co-benefits," are not uniformly measured or validated. Their surplus relies on "the right story" being told and trusted.

The right story is the one routinely produced in news media, positioning carbon crediting Indigenous fire management projects as a "win-win" that have "created jobs in remote and vulnerable communities, improved biodiversity, reinvigorated Indigenous culture" and reduced the frequency of "damaging bushfires."[18] How different individuals narrate this story varies, though often an emphasis is placed on how these projects are "replicating ancient Aboriginal practices" or "combining traditional knowledge of fire management with new technology to protect the landscape." The preferred message is that, as one 2016 headline aptly put it, "tradition leads [the] way on climate."[19] My purpose in foregrounding these dimensions is not to undermine this carbon economy – itself one relatively small formation within the global assemblage of climatic governance – but rather to explore the friction this premium combustive regime generates. As other ethnographic accounts of such economies show, the work of accounting for carbon at different scales and making it perform interscalar leaps produces "friction," in the sense theorized by anthropologist Anna Tsing.[20] Systems of commensuration follow a kind of elemental logic, formatting incoherent, unruly and heterogenous material worlds as units in terms of categories and measures that allow interscalar travel between local, global and other scales. Such formatting produces social energy. It produces dispute, tension, anxiety and disagreement, as a chaotic assemblage of trees, fires, rangers, smoke, gases, scientists, philanthropists, and other idiosyncratic human and nonhuman actors are translated into a coherent story.

Of the ten myths presented for busting that first day of the 2020 Forum, perhaps the most contentious relates to biodiversity. Reading much of what has been written about the topic in scientific journals and news media, it would seem to be axiomatic that these fire management projects are beneficial to this "remote, rugged, biodiversity-rich, and largely unpopulated landscape."[21] This truism is most frequently voiced in statements that such projects are "good for the environment" or "deliver biodiversity benefits."[22] But in the months before the Forum, a group of scientists with ongoing engagements in such projects published a scientific paper suggesting that such benefits were "assumed" rather than demonstrated. As they told it, while it was documented that severe and frequent bushfires harmed biodiversity in various ways, particularly by destroying habitat, and carbon crediting projects reduced the severity and frequency of bushfires, there was little data showing specific benefits to specific species. The paper's esoteric scientific dialect may make this sound like a mild argument, but its implications were more hotly contested. "It's just bullshit," one ecologist told me when I brought it up. "Have you read that paper?" another ecologist asked, as though there was only one article I could have on my mind as we walked to dinner. "It's so racist," he added. Confused, I asked him what he meant. It was racist, he explained, because it implied that Indigenous peoples might be wrong. "You think you're debating, you think you're starting a discussion," he went on, "but you're actually being really offensive and you could kill the fucking market." For him, and others, the mere questioning of benefit was experienced as grievous harm.

Another myth related to the extent to which these projects can be described, as they are when their credits are promoted domestically or internationally, as distinctly traditional or Indigenous. This is a complex matter, but there are various methodological, epistemological and structural factors that both Aboriginal and non-Indigenous people have raised as confounding for these projects. Setting aside questions of how fire management is done – including by dropping napalm capsules from helicopters – the managers and executives of such projects are often white, as are the auditors who check their results, the lawyers who negotiate their contracts, and the scientists who design the systems to calculate the carbon "farmed" and maximize its yield.[23] These factors are well-known within the social networks that converge around the Forum, as people sometimes discuss whether one project is more or less Indigenous in its governance structure or the cultural standing of its Aboriginal practitioners. But those who draw outside attention to these factors are often addressed with hostility. This was demonstrated repeatedly in relation to the work of an anthropologist who published research, on the basis of ethnographic fieldwork, suggesting that a carbon project conformed to the demands of non-Indigenous experts and systems and impeded local Aboriginal peoples' decision-making. Again, the relatively mild assertion that an introduced economy might have a disciplining effect was received as a betrayal by many. As one white fire scientist told me, the implication "that we are subverting

Aboriginal peoples' autonomy or independence somehow is stupid." It is also unhelpful, he added, as "it weakens the potential to tell a unified story" to investors.

A third myth, that projects lead to people burning areas too frequently, signals the tensions surrounding the formatting of combustion itself. The sanctioned carbon abatement methodology codes fires and their emissions in a binary fashion. Throughout the year, somewhere above Australia's tropical savanna, an infrared sensor on a satellite is watching for the exothermic signature of bushfire, mapping each fire on a scale of 1 square kilometre per pixel. If a pixel burns before the first day of August in a given year, the whole pixel is coded as an Early Dry Season (EDS) fire, but if it is after this date it is coded as a Late Dry Season (LDS) fire. EDS and LDS are each coded as having singular emissions values, the former much lower than the latter, based on aggregated empirical measures. The rule has become that, as one ecologist said to me, "EDS is good and LDS is bad," with upward shifts in EDS numbers reported publicly as uncomplicated signs of success. But proponents are well aware that the emissions of fires are driven by their intensity, itself a product of seasonality and climate, and not calendar dates. A fire on the last day of July can be far more intense, and therefore produce more emissions, than a fire on the first day of August. Similarly, all are aware that fires do not burn in discrete pixels, and that processing a fire that burns a fraction of a pixel identically to a fire that burns all of it is imprecise. Nonetheless, the simplicity of the system is held by many as a virtue, even if it allows exploitation at the margins or means that proponents burn areas more frequently or intensely than is ecologically optimal. When journalists have, very occasionally, reported concerns about the seeming annual rush to light up the land in the last weeks of July, proponents have again closed ranks. The possibility of ambiguous media coverage means, as a white project executive told me, many only "do media" when they can control it.

What I have left out of my own attempt to narrate this economy is that there are serious questions as to whether or not it could survive without a premium story and its associated economic value. Since 2013, if not before, the scalar politics of Australian national climate policy has been adverse (if not hostile) to the interests of these projects and their proponents. The government of this emissions-intensive nation has sought to simultaneously quell domestic economic fears and meet its international emissions targets by buying carbon abatements at as low a cost and quality as possible. It has excluded other governments from its market to reduce competition and used reverse auctions through the Emissions Reduction Fund – recently rebranded as the Climate

Solutions Fund – to suppress prices, keeping them at levels such that the profit margin for many Indigenous carbon projects is a tenuous 5 per cent to 10 per cent per credit. For these reasons, it is understandable that many project proponents see plans to measure and verify their purported premium values, or reform their calculative methods, as potential impositions of unaffordable costs rather than routine scientific critiques. "You have to understand it is not a climate change program," one ecologist told me, "it's an Indigenous jobs program." To many Indigenous and non-Indigenous proponents, the material impact of these projects on carbon balances is of secondary import, or sometimes of no importance at all, compared to the local employment and income their premium value helps generate. Holding such projects together – themselves constituents of one formation within carbon's changing elemental economies – requires a series of compelling and competing narratives about savanna fires, and their overlapping commensurabilities and incommensurabilities. Holding these projects together requires contests around the right story. Rather than errors, or reducible imperfections, we might therefore view the arguments, swearing, bluster, angry emails and other heated and frictive interactions of my interlocutors as constitutive of elemental exchange.[24]

Spending time in these combustive contexts, it is easy to become absorbed in their everyday politics, or what MacKenzie calls their "subpolitics," and lose sight of the politics of commensuration. From a distance, one could easily fixate upon the familiar forms of domination that this economy arguably maintains and reinscribes in its production of tradeable carbon, enacting "carbon colonialism" by pressuring Indigenous peoples to use and manage their land in certain regulated ways for the convenience of a white majority already enriched through the mass dispossession of Indigenous peoples. Following Lohmann and others, in their glowing public image as forms of "climate change action," these carbon projects also arguably distract from the more substantial changes in emissions required to avoid catastrophic climate change. Indigenous carbon credit projects figure prominently in the public reporting of oil and gas multinationals such as ConocoPhillips and INPEX though these projects collectively represent less than a percentage of these multinationals' direct or downstream emissions. Indigenous carbon credit projects also represent a fraction of the total market and a smaller fraction of the carbon credit portfolios of private and public institutions.

But such political critiques are not simply unsympathetic or unreasonable but also unrealistic, in the sense that they are disengaged from the social realities of the actors that constitute these markets at different scales. Up close, in the hot and flammable savanna grasslands, I have found that arguments about coercive governance and the logics of carbon colonialism are typically met with confusion, disregard or, sometimes, disgust. The non-Indigenous actors engaged in Indigenous carbon projects, for example, do not pretend to be revolutionaries against hegemony. However, they do see their work as locally transformational, manipulating the possibilities of marketized carbon – the contingencies of global atmospheric relations – to support the individuals and social worlds they value. As a universalizing measure of life in the Anthropocene, carbon is both a device for formatting and foreclosing social worlds and, at the same time, maintaining and manifesting their possibilities.[25]

NOTES

1 See Dionne Walsh, Jeremy Russell-Smith, and Robyn Cowley, "Fire and Carbon Management in a Diversified Rangelands Economy: Research, Policy and Implementation Challenges for Northern Australia," *The Rangeland Journal* 36, no. 4 (2014).

2 DISER, *Northern Australia Fire Management with the Emissions Reduction Fund* (Deptartment of Industry, Science, Energy and Resources, 2020).

3 Jerome Whitington, "Carbon as a Metric of the Human," *PoLAR: Political and Legal Anthropology Review* 39, no. 1 (2016).

4 Donald MacKenzie, "Making Things the Same: Gases, Emission Rights and the Politics of Carbon Markets," *Accounting, Organizations and Society* 34, no. 3 (2009).

5 Eva Lövbrand and Johannes Stripple, "The Climate as Political Space: On the Territorialisation of the Global Carbon Cycle," *Review of International Studies* 32, no. 2 (2006).

6 Ingmar Lippert, "Environment as Datascape: Enacting Emission Realities in Corporate Carbon Accounting," *Geoforum* 66 (2015).

7 MacKenzie, "Making Things," 453.

8 Andrew Baldwin, "Carbon Nullius and Racial Rule: Race, Nature and the Cultural Politics of Forest Carbon in Canada," *Antipode* 41, no. 2 (2009); Heather Lovell and Diana Liverman, "Understanding Carbon Offset Technologies," *New Political Economy* 15, no. 2 (2010).

9 Heidi Bachram, "Climate Fraud and Carbon Colonialism: The New Trade in Greenhouse Gases," *Capitalism Nature Socialism* 15, no. 4 (2004): 15; Anders Blok, "Clash of the Eco-Sciences: Carbon Marketization, Environmental NGOs and Performativity as Politics," *Economy and Society* 40, no. 3 (2011); Baldwin, "Carbon Nullius."

10 Janelle Knox-Hayes, "Constructing Carbon Market Spacetime: Climate Change and the Onset of Neo-modernity," *Annals of the Association of American Geographers* 100, no. 4 (2010).

11 Larry Lohmann, "Neoliberalism and the Calculable World: The Rise of Carbon Trading," in *The Rise and Fall of Neoliberalism: The Collapse of an Economic Order?*, ed. K. Birch and V. Mykhnenko (London: ZED Books, 2010).

12 David M Lansing, "Performing Carbon's Materiality: The Production of Carbon Offsets and the Framing of Exchange," *Environment and Planning A* 44, no. 1 (2012): 218.

13 Pablo Jaramillo, "Sites, Funds and Spheres of Exchange in a Clean Development Mechanism Project," *Journal of Cultural Economy* 11, no. 4 (2018); Whitington, "Carbon as a Metric."

14 Véra Ehrenstein, "The Friction of the Mundane: On the Problematic Marketization of the Carbon Stored by Trees in the Tropics," *Journal of Cultural Economy* 11, no. 5 (2018); Jerome Whitington, "The Prey of Uncertainty: Climate Change as Opportunity," *Ephemera: Theory & Politics in Organization* 12 (2012).

15 Gabrielle Hecht, "Interscalar Vehicles for an African Anthropocene: On Waste, Temporality, and Violence," *Cultural Anthropology* 33, no. 1 (2018): 135.

16 Hecht, "Interscalar Vehicles," 113, 35.

17 Cole Latimer, "Indigenous Group Launches Nation's First Private Carbon Trading Fund," *The Age*, 4 May 2018; Sue Jackson et al., "Cultures of Carbon and the Logic of Care: The Possibilities for Carbon Enrichment and Its Cultural Signature," *Annals of the American Association of Geographers* 107, no. 4 (2017).

18 Sam Johnston, "Indigenous Innovation Could Save a Billion Tonnes of Greenhouse Gases," *The Conversation*, 21 April 2016; Emma Masters, "Gas Giant Inpex Multi-Million Dollar Greenhouse Gas Offset Program Under Fire," *ABC News*, 21 July 2015.

19 Kristy O'Brien and Neda Vanovac, "Arnhem Land Rangers Blending Traditional and High-Tech Methods to Earn Carbon Credits," *ABC News*, 10 September 2017; Michael Slezak, "'The Idea Is Coming of Age': Indigenous Australians Take Carbon Farming to Canada," *The Guardian*, 23 July 2017; Danaella Wivell, "Tradition Leads Way on Climate," *Cairns Post*, 25 October 2016.

20 Anna Lowenhaupt Tsing, *Friction: An Ethnography of Global Connection* (Princeton: Princeton University Press, 2005); Wim Carton, "Rendering Local: The Politics of Differential Knowledge in Carbon Offset Governance," *Annals of the American Association of Geographers* (2020); Ehrenstein, "The Friction of the Mundane."

21 Jeremy Russell-Smith et al., "Managing Fire Regimes in North Australian Savannas: Applying Aboriginal Approaches to Contemporary Global Problems," *Frontiers in Ecology and the Environment* 11, no. s1 (2013).

22 Jeremy Russell-Smith and Peter J. Whitehead, "Reimagining Fire Management in Fire-Prone Northern Australia," in *Carbon Accounting and Savanna Fire Management*, ed. B.P. Murphy, et al. (Collingwood, Vic: CSIRO Publishing, 2015).

23 For more on these types of relationships in Indigenous institutions and organizations, see Philip Batty, "Private Politics, Public Strategies: White Advisers and Their Aboriginal Subjects," *Oceania* 75, no. 3 (2005).

24 Carton, "Rendering Local."

25 Whitington, "Carbon as a Metric," 57.

3 CEMENT

eli elinoff

Cement might be the very first anthropogenic element. By crushing, mixing, and cooking stones rich in calcite ($CaCO_3$), humans created a new form of stone that radically remade the planet. Cement and its cognate, concrete, have redirected fluvial pathways, levelled mountains, extended shorelines, excavated sea beds, and produced new landmasses.[1] These materials have so profoundly reshaped the planet's topography that they have become a key marker of the earth's most recent geological layer: the active, anthropogenic stratigraphy called the "archaeosphere."[2]

Beyond its earthbound effects, cement production has transformed our atmospheres. The production of cement is energy intensive and, therefore, carbon intensive.[3] Cement is responsible for between 5 per cent and 8 per cent of the world's annual carbon emissions.[4] In addition to this, the production of dust during the processing of limestone extraction and concrete construction are responsible for forms of both large and small particle pollution that lodges in the lungs of quarry workers, construction crews, and urbanites.

It would be impossible to imagine modernity without anthropogenic stone. The material is essential to the physical construction of most modern infrastructures and thus to the spatial and logistical synchronization of much of the planet. Ports, highways, and dams weave together new relations across space and time, accelerating interaction by linking sites of extraction, production, distribution, and consumption in new ways.[5] Logistics, geographer Deb Cowen notes, are closely related to currents of militarization that structure most twentieth and twenty-first century geopolitics.[6] The inextricable relationship between war and the critical infrastructures of modernity reveal how cement and concrete shape the terms of life and death in the modern world.[7]

Concrete construction is fast, cheap, and based on materials that, for the most part, remain widely available.[8] It can be deployed in both extremely complex built forms but also rudimentary ones.[9] Stripped of adornment, smooth béton brut – raw concrete – offered modernist architects like Le Corbusier or Oscar Niemeyer a means of designing structures whose mass and use of space was aimed at forging a radical break with the past through the literal construction of a new future.[10] In this, the material's qualities – smoothness, hard edged, functional – are also critical to the aesthetic possibilities of modernity. Indeed, the temporal rupture posited by modernism itself was deeply connected to concrete's perceived ability to stand outside of history. These visions became central to both imperial and nationalist visions of development.[11]

For the modernists, this was a progressive future, but, as Elaine Gan and colleagues remind us, the relationship between modernity and the Anthropocene is also "suffused with the bad death ghosts."[12] Every concrete building, road, highway, and housing estate is composed of mined mountains themselves made up of the fossils of long-dead marine life.[13] The loss of habitats and fossils foregrounds the connections between the biological and geological loss in cement productions and the forms of violence associated with militarism, colonialism, and capitalism that shape most construction sites in some way. Andrew Alan Johnson reminds us that these deaths do not always rest easily inside finished structures, but instead re-emerge in moments of haunted capitalist failure.[14]

Despite its status as an elemental part of modernity's infrastructures, cement has remained absent from most discussions of planetary environmental transformation. Attention to cement – both its production and deployment – reveals the *concrescence* of social, technical, ecological, and economic forms in the production of geopolitical speed.[15] In Thailand, the geographic focus of this essay, the emergence of cement altered the country's topographies and atmospheres. Cement and concrete resynced territorial relations between citizens and acted as a means of militarized exchange, solidifying and corroding international orders simultaneously. Cement and concrete reshaped relations between economies, ecologies, and polities in ways that reveal precisely how essential human-made stone is to the geopolitics of planetary environmental change.

Anthropogenic Stone

Cement is produced by cooking the calcium-rich fossilized remains of marine organisms. The chemical reactions required to create modern Portland cement take place at 1450 degrees Celsius.[16] As "raw meal" – composed primarily of limestone and lesser amounts of clay, chalk, and other minerals rich in calcite – comes to temperature, the minerals begin to break down and transform. Their stored water is released, calcium carbonate becomes calcium oxide, and a host of other chemical reactions take place that transform the molecular bonds inside the materials. As the minerals cool they become "clinker," stony grey lumps that, when ground and mixed with gypsum, become cement.[17]

The elemental process of cement making has been evident for as long as the material has existed. Remarking on ancient forms of cement, the Roman Architect Vitruvius used the "five elements theory" to explain the chemical reaction I described above. For Vitruvius, cement results from the application of fire to stone which weakens the stone. The later application of water to this diminished stone helps it "recover its vigor."[18] Rereading Vitruvius in the context of Colombia, anthropologist Michael Taussig notes how the cooking of limestone for cement is enmeshed at both a material and economic level with the making of cocaine, arguing that humans generate speed through the production of both substances.[19] As Neale, Addison, and Phan point out in the introduction to this volume, "an elemental perspective is one that tries to locate an analytic starting point, and then traces the cacophony of reactions that unfold from there." This call to the elemental, read alongside Taussig's insights into cement's speed, motivates me to suggest that the biomass of disappeared seas, the extractive force of limestone mining, and the geological alchemy of cement production are important starting points for understanding anthropogenic geologies and their geopolitics.

As geographer Kimberly Peters argues, "It is necessary to start thinking beyond a world where geopolitics is a process wrought through the 'earth' or terra alone."[20] Instead, she asks us to consider the ways in which elements mix, mingle, blend, and act to stabilize and destabilize various topographies and even atmospheres.[21] For example, in the process of making clinker, carbon is released in two different ways: First, carbon stored in the mineral form of calcium carbonate in limestone or chalk is released as it becomes clinker. Second, clinker production requires intense heat produced by burning fossil fuels.[22] Cement production thus enrols other elements – such as coal and fuel oil – in the making novel geologies and atmospheres. The expanding uses of cement are, thus, an elemental, even if an obscured part, of what Timothy Mitchell calls "Carbon Democracy."[23]

Cheap, steady supplies of domestic cement are essential for the mass construction that drives much development. Indeed, cement manufacturing is usually an early accomplishment in modernizing countries.[24] This means two things: first, that stone extraction remains a critical, underexplored aspect of modernization projects. Second, cement is always associated with secondary forms of extraction, of often rudimentary fossil fuels that are necessary to keep kilns operating. Cheap, dirty, and hot, coal and fuel oil tend to be the primary means through which cement kilns are heated. This interdependence of cement and cheap fossil fuels extends the environmental reach of the material as its production requires other forms of extraction.[25]

Cement, taken in this sense, is not a stable, uniform actor deployed to bring the Earth under anthropogenic management. Rather, it is an unsteady, dynamic, always unfinished outcome of material, ecological, and political processes.[26] It is *geopolitical*, in the richest, most material sense of that word. Imaginaries of cement reflect illusory dreams of an organized stable, secure modernity. However, as an unstable anthropogenic element, cement's processes can be turbulent, fragile, and violent. It both generates power and corrodes it. Cement-making entails socio-political as well as environmental processes. The extraction of stone from the land requires enclosures, dispossession, and the displacement of humans, non-humans, and geomateriality. Cement production demands heat. Construction generates further dispossession as land is transformed into property. It also requires the deployment of regimes of skilled and unskilled labour and the application of new forms of expertise.[27]

Attention to cement's elemental status thus helps situate the fossils of long-dead marine organisms deeply within our contemporary political and environmental moment. Cement's relevance as a key anthropogenic element that mixes and extends a wide range of environmental impacts exceeds its carbon metrics. The modern era of cement production is a critical component in the making of a geopolitics in which stone is not a secondary part of political ecological struggle but, rather, fundamental. These effects become visible when we trace the material's origins and trajectories in specific places – such as contemporary Thailand.

Cementing Siam

A Thai friend of mine once called Bangkok a "clogged artery" in the heart of Thailand's Chao Phraya delta. This metaphor was an essential one during the catastrophic floods of 2011, which cost the country nearly US$45 billion. News reports during the calamity used similar circulatory metaphors, calling the flood as a "cerebrovascular accident." The floods left the heterogenous interface between the city's hardened surfaces and soft delta-landscapes in Bangkok's peripheries inundated for months as politicized upcountry water management encountered a late rainy season and astronomical high tides. Pundits used the metaphor of a stroke to justify the need to let the urban periphery drown to save Bangkok's urban core.

Despite the fantasies of fixity conjured by cement and concrete, the metaphor of the clogged artery draws attention to the century of material churn that spread the region's Permian-era Limestone Karst across the Chao Phraya River delta.[28] As the steep and twisted karsts were levelled, they re-emerged in the form of industrial estates, shopping malls, housing estates, paved roads, and highways. To walk in Bangkok is to weave through stacks of cement bags and dodge teams of construction workers, typically from Myanmar, Cambodia, or northeastern Thailand, simultaneously jackhammering up existing pavement, while also mixing cement, sand, and water in an effort to smooth over cracking streets. The din of construction is constant in most parts of the city. This is the sound of earth being relocated.

As a delta city, building Bangkok involved filling in its vast wetlands. To build anything requires filling and redistributing earth, which is then covered over in a crusty shell to accommodate new houses and industrial estates. Yet, delta ecologies are stubborn, multi-species affairs.[29] Their tendency to flood does not end merely because they have been filled in. City streets are raised, flood walls are built, and new pipes driven through soft clay foundations, all in the name of managing the relation between hard and liquid forms.

In the topographic story of cement in Thailand, 1915 is a crucial turning point in Thai environmental history. In that year, two years after being founded as joint-partnership between the Siamese monarchs and Danish engineers, the Siam Cement Company began producing the first domestic bags of cement in the country from its kilns.

At roughly the same time, canal construction in Bangkok ceased, and the city, which was previously organized around aquatic transport, became increasingly terrestrial.[30] Roads, alleyways, highways, and, eventually, elevated and subterranean mass transit were constructed along these canal routes. More often, these land-based forms of travel replaced the waterways entirely. New modes of automobility supplanted aquatic mode of travels. The shift towards terrestrial relations remade the Chao Phraya river's topography and its socio-environmental relations, which intimately linked people to the seasonally fluctuating water through deep water rice cultivation and fishing.[31]

Siam Cement was created to provide an indigenous source of cement. The company imported technology and new kinds of expertise from Denmark. The expansion of the industry entailed the mapping of new material landscapes as well. The company's first marl (chalk) pits were located roughly 170 kilometres outside of Bangkok, but by the middle of the 1920s new marl pits were found in a region called Saraburi, about half that distance from the capital.[32] These marl pits laid the foundation for the geographies of extraction and production associated with the contemporary cement industry; Saraburi is now the undisputed centre of limestone extraction in the country. These pits also set up new possibilities for infrastructural and military relations in the Cold War period, which I describe in the next section.

The construction of road and rail expanded the uses of cement, putting them towards the production of new infrastructures and growing cities. However, as architectural scholar Lawrence Chua argues, cement's importance was more than utilitarian: reinforced concrete bridges, shophouses, temples, and palaces generated a nationalist aesthetic for the country that evoked ideas of masculine heroism, militarism, racial homogeneity, and national sacrifice. As Chua notes, reinforced concrete and cement were key to creating ideas of citizenship for the country as it sought to assert itself as a modern state.[33] Thailand's nationalist aesthetics also generated new construction techniques as heavier structures required that concrete posts be driven deep into the soft soil known as "Bangkok clay."[34] While the materials generated new political aesthetics they also had geological and architectonic implications. The convergence of these things reveals precisely how, from its outset, the creation of a Thai state was a deeply geopolitical process.

From Cold War Infrastructures to Capitalist Real Estate Booms

Beginning in the middle of the 1950s, the scale of transformation of the country's rivers expanded. New dams in the Chao Phraya river basin remade the delta's hydrologic landscape, reorganizing people and space. The dams simultaneously made new kinds of farmland irrigation possible while also producing new grounds for urbanized industrialization in the delta.[35] Like the founding of Siam Cement, these ecological transformations were produced out of webs of materials and political relations. As anthropologist Jakkrit Sangkhamanee describes, the construction of the Chao Phraya Dam, the country's first large dam, required a massive social and technical coordination between US and Thai engineers who collaborated on the design and construction of this new infrastructure.[36]

The country's second large dam, the Bhumipol Dam, named after King Bhumipol Adulyadej, proceeded through similar international collaborations. The dam, which, at the time, was touted as the largest in Asia, required upwards of 300,000 tons of cement. This heavy demand forced the Thai government to create a new, state-run cement company – Jalaprathan Cement.[37] Founded in 1956 by the Royal Irrigation Department, the company began construction on a cement factory in Takhli province in 1958. The site was also not far from deposits of lignite coal, which eventually transformed into one of the largest open pit mines in mainland Southeast Asia. Further, the Jalaphrathan factory was located just a few kilometres from a Thai air force base built in the early 1950s. That base would serve as a platform for covert American missions to Tibet during its war with China and, later, as a key base of operations for US bombing in Laos, Cambodia, and Vietnam during the Southeast Asian wars.[38] These interconnected geographies of development, industrial extraction, militarization, and environmental change suggest how cement tied both regional and global geopolitics to changing local ecologies in important new ways.

The US military was a major customer of Siam Cement during its wars in Southeast Asia, purchasing large quantities of the material to build new roads and military installations. Historian of science John DiMoia characterizes these massive infrastructure projects, which brought together American capital, Korean engineers, and Thai labour as "security roads."[39] Aimed at facilitating troop movements, opening up potentially dissident regions, and acting as launching points for bombing raids, these militarized infrastructures relied upon massive quantities of locally produced cement.[40] In addition to roads, the US government built several airbases, which continue to operate today as grounds for the Royal Thai Airforce. Although the US stopped using these bases in the 1970s, they were recently associated with revelations of American "Black Sites" used to undertake interrogations in the War on Terror. These rumours emphasize the long and complex legacies of military constructions in the region.[41]

Cold War militarization generated many of the essential infrastructures that facilitated Thailand's second economic boom, which began in the 1980s, transforming the country into Asia's fifth newly industrialized country. Bangkok's population nearly tripled during the Vietnam war era. This population growth was in part due to the growth of the industrial sector, but also due to urbanization related to the large American military presence. New bars, restaurants, and hotels cropped up to serve American forces during the period.[42] Urbanization and economic transformation further increased cement production, as new modes of dwelling and labouring took hold throughout the urbanizing country.

In the 1980s Bangkok grew even faster. New high-rise apartment complexes and skyscrapers sprouted across the city. In the provinces, houses, which were previously constructed of timber, open to the elements, and often raised to accommodate periodic floodwaters, were increasingly built on cement foundations. Migrants often sent money home to pay for concrete blocks to fill in the open designs, which accommodated floods and allowed for cool breezes to pass through.[43]

From the mid-1980s until the Asian economic meltdown in 1997, the real estate market in Bangkok was so hot that there were supply shortages of cement.[44] During this period, the country's productive capacity expanded exponentially, with a range of new industrial players entering the market.

Thailand's cement production reached its first peak shortly before the 1997 economic crash at over 40 million tons of cement per year. Factory workers employed at Siam Cement's Khaeng Khoi plant at the time told me that there were lines of cement trucks extending for kilometres out of the industrial facility during this period. These lines remind us that both militarized infrastructures and, later, real estate expansion were fuelled by the mining of tons of stone. These workers also recalled that after the economy crashed there was a major glut of cement. Although the housing market had cooled, hot kilns continued to churn the material at the same rate as before. Supply well outstripped demand. Thai companies began exporting the material for the first time. Some former plant employees told me that companies used mislabelled bags to ship the material anywhere that would take it. As Thailand's construction sector ground to a halt, empty, half-built ghost towers became haunted emblems of the region's collapse and fed an emerging mistrust of the promise of capitalist modernity and its promises of progress. The obdurate, half-built structures remind Thais that the material's solidity often betrays its own promises.[45]

Modernity Inundated

In 2011 Bangkok was hit by historic floods. The concretization of the river delta had transformed it into a tangle of amphibious infrastructures, some of which sped the flow of water while others blocked it.[46] Instead of simply being a result of high rainfall, scholars have demonstrated that the 2011 flood occurred as a result of the intersection between politicized water management, a late rainy season, and an astronomical high tide.[47] If the dawn of the Anthropocene in Thailand appears around 1915 with the founding of the cement company and the shift towards terrestrial living, the flood offered a glimpse of futures yet to come.[48]

After the floods, the country's hard urban surfaces moved to the fore of people's environmental consciousnesses. Architects and planners proposed new designs for flooded cities. Protestors mobilized to prevent further damming upstream. Government offices attempted to create a massive coordinated flood infrastructure. Civic groups have sprung up to debate the city's scant green space and to advocate for better care of existing trees. Across the country, speculative imaginaries of "living with flood" have enabled people opportunities to think through the situated effects of climate change in the present.[49]

Few of these ambitious amphibious urban projects have come to pass. Instead, the 2014 military coup led to a boom in traditional infrastructure spending, much of which came with promised linkages with China's Belt-Road Initiative. Housing developments, urban infrastructures, rail projects, and special economic zones across the country reflected this new building boom. Even with insights about the way the 2011 floods were exacerbated by urban growth into the city's wetlands, delta ecologies continue to be filled in at an unprecedented rate. Although environmentalists scuttled a more comprehensive flood plan due to its reliance on large dams, local governments have constructed a patchwork of floodwalls, dikes, and canals, mainly through financial disbursements to fragmented authorities working in tandem with local governments.[50] These splintered flood infrastructures resulted in spatial fragmentation as previously continuous tracts of land were divided by raised roads and walls. They have also repartitioned future flood risks; some places will remain dry while others will be inundated.[51]

When they are read alongside the post-2014 military coup construction boom, these projects highlight the way cement and concrete remain essential drivers of the economy during difficult times. They also show how fantasies of smoother, more secure futures produce dust and topographic churn and political struggle in the present. Rather than evoking a sense of certainty, cement constructions after the 2011 floods have drawn the breakneck speed of Thailand's twentieth century modernization project into relief against a tumultuous future in which the delta's encrusted riparian topography remains at the core of environmental questions over the changing planetary ecologies.

At the Speed of Earth

Though neither the economic nor political histories in Thailand have been linear, the rapid expansion of cemented environments has generally increased across the twentieth and twenty-first centuries. Cement and concrete bring with them a host of questions about the geopolitics of anthropogenesis and its elements. My brief narrative of cement and environmental transformation in Thailand emphasizes how the application of cement and concrete have been closely associated with an accelerating curve of uncertain environmental changes. These changes can be localized in terms of a square metre of sidewalk or expanded to include the destruction of mountainsides for limestone, the expansion of coal mining to fire kilns, the re-sculpting of river basins and deltas

through dams and hardscape irrigation, the creation of flood infrastructures that require human management, and the growing concern
around atmospheric carbon. Following Nicholas D'Avella's insight that
cement is "never alone," this history shows how it is always in relation
with social, environmental, and political worlds in the making.[52]

Gabrielle Hecht notes that attention to certain objects can overcome
the dilemmas posed by the Anthropocene by helping us traverse across
unbridgeable scales of space and time.[53] Attention to cement's elemental
status might do this and more: close analysis of the material's uses, imaginaries, and its geopolitics also help us make sense of the geopolitical and
socio-environmental dynamics the material generates as it moves. In this
we might return to Taussig's insights to consider how cement is an elemental part of the generation of the speed of contemporary life. As Paul
Virilio notes, the street is a "producer of speed" of all sorts.[54] Similarly,
cement and concrete are tied to speed. They were critical to both imagining the developmental rupture and generating it. Concrete construction propelled local economies forward, sped up growth, and resulted in
breakneck territorial transformation. In the case of Thailand, the material allowed the country to "catch up" with global developmental norms
but radically altered its ecologies in the process. Indeed, in large parts of
the world these materials continue to sustain the geological speed that
underrides global economy.

My point here is not to overstate the role of cement in global development or to reduce the causation of our contemporary environmental
catastrophe to its production; this would be inaccurate. Rather, Thailand's histories of cement and concrete, like the history of those materials elsewhere, reveal the profound role these materials play in sustaining
the rapid pace and the uneven, unpredictable character and violent politics of life in Anthropocene. In this sense, cement is elemental to these
times. We cannot imagine this new geological temporality, in which human life now presses us past speed of earth, without it.

Cement is thus elemental to human-generated geological speed.
Thinking with cement allows us to consider the varied kinds of politics
that have been central to and have flowed from this achievement. This
becomes especially clear when tracing cement and concrete through
specific contexts at specific times. As I have shown in this chapter, following Thai cement through the country's histories of modernization,
militarization, and capitalist development reveals precisely the kinds
of power that have often flowed alongside and through the geological
acceleration occurring as wet concrete hardens in the varied forms of
the built environment.[55] Following cement in the Anthropocene thus
compels us to consider the multiple socio-political and environmental

processes that generated the material's histories and how those same processes are transforming in the present, speeding us towards new environmental futures emerging right beneath our feet.

NOTES

1 Concrete is a compound material that contains cement, water, sand, and aggregates. In this chapter, I use cement to refer specifically to that material and concrete to refer to this wider compound. In common speech these words are often used interchangeably, but here I mobilize them slightly differently to emphasize cement's elemental status while also to highlight concrete's capacity to extend the impacts of cement through more complex material convergences. See also Nicholas D'Avella, *Concrete Dreams: Practice, Value, and Built Environments in Buenos Aires* (Durham: Duke University Press, 2020), 6.

2 Matt Edgeworth. "More than Just a Record: Active Ecological Effects of Archaeological Strata," in *Historical Archaeology and the Environment*, eds. M.A.T. De Souza and D.M. Costa (Cham: Springer, 2018), 19–40. Emily Elhacham et al. estimate that human-made mass now exceeds all living biomass. Their estimates add credence to Edgeworth's arguments. Emily Elacham et al., "Global Human-Made Mass Exceeds All Living Biomass," *Nature* 588 (2020): 442–4.

3 Usually, cement kilns, which must reach sustain temperatures of 1400 degrees Celsius, are fired by coal or fuel oil.

4 Ernest Worrell et al., "Carbon Dioxide Emissions from the Global Cement Industry," *Annual Review of Energy and Environment*. 26, no. 1 (2001): 303–29. See also Robbie Andrew, "Global CO_2 Emissions from Cement Production 1928–2018," *Earth Systems Scientific Data* 11 (2019): 1675–710.

5 Penelope Harvey, "Cementing Relations: The Materiality of Roads and Public Spaces in Provincial Peru," *Social Analysis* 53, no. 2 (2010): 28–46.

6 Deborah Cowen, *The Deadly Life of Logistics: Mapping Violence in Global Trade.* (Minneapolis: University of Minnesota Press, 2014).

7 For example, Greg Dvorak, *Coral and Concrete: Remembering Kwajalein Atoll between Japan, America, and the Marshall Islands* (Honolulu: University of Hawaii Press, 2018); Kali Rubaii, "Concrete and Livability in Occupied Palestine," *Engagement: A Blog,* Anthropology and Environment Society, 2016, https://aesengagement.wordpress.com/2016/09/20/concrete-and-livability-in-occupied-palestine/.

8 Recent global shortages of sand are an exception here.

9 Adrian Forty, *Concrete and Culture: A Material History* (London: Reaktion, 2012).

10 For example, James Holston, *The Modernist City: An Anthropological Critique of Brasilia* (Chicago: University of Chicago Press, 1989).

11 For example, Diana Martinez, *Concrete Colonialism: Architecture, Infrastructure, Urbanism, and the American Colonization of the Philippines* (PhD diss., Columbia University, 2017); James Holston, *The Modernist City: An Anthropological Critique of Brasilia* (Chicago: University of Chicago Press, 1989).

12 Elaine Gan et. al., "Introduction: Haunted Landscapes of the Anthropocene," in *Arts of Living on a Damaged Planet: Ghosts and Monsters of the Anthropocene*, eds. Anna Lowenhaupt Tsing et al. (Minneapolis: University of Minnesota Press, 2017), G7.

13 I thank both Heid Jerstad and Cristián Simonetti for highlighting why cement's biological roots matter for our thinking about the material's contemporary environmental impacts. See Heid Jerstad, "Kinship," in *The Social Properties of Concrete*, eds. E. Elinoff and K. Rubaii (punctum books: Earth, forthcoming). See Cristián Simonetti, "Time," in *The Social Properties of Concrete*, eds. E. Elinoff and K. Rubaii (punctum books: Earth, forthcoming).

14 Andrew Alan Johnson, *Ghosts of the New City: Spirits, Urbanity, and the Ruins of Progress in Chiang Mai* (Honolulu: University of Hawaii Press, 2014), 30.

15 See Eli Elinoff and Kali Rubaii, "Introduction: The Social Properties of Concrete," in *The Social Properties of Concrete*, eds. E. Elinoff and K. Rubaii (punctum books: Earth, forthcoming).

16 The history of Portland cement, which is also called normal cement, is complex. Although it was patented in England by Joseph Aspdin in 1822, there are a range of competing stories about the material's development. The manufacturing process and material composition are key differences between the modern Portland cement and so-called Roman cement, which has existed for thousands of years.

17 Michael Gibbs, et al., "CO_2 emissions from Cement Production," in *Good Practice Guidance and Uncertainty Management in National Greenhouse Gas Inventories*, 2000. https://www.ipcc-nggip.iges.or.jp/public/gp/bgp/3_1 _Cement_Production.pdf.

18 Quoted in Stephen L. Sass, *The Substance of Civilization: Materials and Human History from the Stone Age to the Age of Silicon* (New York, Arcade Publishing, 1998), 130.

19 Michael Taussig, *My Cocaine Museum* (Chicago: University of Chicago Press, 2004), 161.

20 Kimberley Peters, "Elements," in *Territory Beyond Terra*, eds. K. Peters et al. (Lanham, ME: Rowman & Littlefield International, 2018): 18.

21 Kimberley Peters, "Elements," 18.

22 Andrew, "Global CO_2 Emissions," 1675.

23 Cement production is a key part of the larger histories of energy extraction and therefore atmospheric carbon. However, the politics that unfold from cement production must also be read in their own context in relation to specific politics of emplaced environmental harm.

24 For example, Emily Brownell, *Gone to Ground: A History of Environment and Infrastructure in Dar es Salaam* (Pittsburgh: University of Pittsburgh Press, 2020).

25 The cement industry has undertaken efforts to green its images and practices, launching The Cement Sustainability Initiative in 1999. However, the geographies of cement production are politically and economically complex. Because the industry is capital intensive, technological advancements in mature production centres may or may not reach emerging centres. Moreover, emerging sites of production are spaces in which basic technologies, lax regulatory enforcement and low legal protection play key roles in shaping production processes. What sustainability might mean in any one context depends not only on the sorts of technology being used but also on the strength and reach of regulatory law, both of which depend deeply on their political and economic contexts.

26 Elinoff and Rubaii, *Social Properties*, forthcoming.

27 Penelope Harvey, "Cementing Relations: The Materliality of Roads and Public Spaces in Provincial Peru," *Social Analysis* 54, no. 2 (June 2010): 28–46.

28 Sur suggest concrete churns earths and histories in its deployment. Malini Sur, "Churning," in *The Social Properties of Concrete*, eds. E. Elinoff and K. Rubaii (punctum books: Earth, forthcoming).

29 Atsuro Morta, "Multispecies Infrastrcture: Infrastructural Inversion and Involutionary Enganglements in the Chao Phraya Delta, Thailand," *Ethnos* 8, no. 24 (2017): 738–57.

30 Although road construction began in earnest in the middle of the nineteenth century, it wasn't until 1915 that new canal construction stopped. See Porphant Ouyyanont, "Physical and Economic Change in Bangkok, 1851–1925," *Japanese Journal of Southeast Asian Studies* 36, no. 4 (1999): 456.

31 Brian McGrath, et al., "Bangkok's Distributary Waterscape Urbanism," *Village in the City: Asian Variations of Urbanisms of Inclusion* (Chicago, IL: Park Books–UFO: Explorations of Urbanism, 2013).

32 Siam Cement Group, *100 Years of Innovations for Sustainability* (Bangkok: SCG, 2013), 26.

33 Lawrence Chua, "The City in the City: Race, Nationalism, and Architecture in Early Twentieth Century Bangkok," *Journal of Urban History* 40, no. 5 (2014): 933–58.

34 Koompong Noobanjong, *The Aesthetics of Power: Architecture, Modernity, and Identity from Siam to Thailand* (Bangkok: White Lotus Press 2013).

35 Atsuro Morita, "Infastructuring Amphibious Space: The Interplay of Aquatic and Terrestrial Infrastructures in the Chao Phraya Delta in Thailand," *Science as Culture* 25, no. 1 (2016): 117–40.

36 Jakkrit Sangkhamanee, "Infrastructure in the Making: The Chao Phraya Dam and the Dance of Agency," *TRaNS: Trans-Regional and -National Studies of Southeast Asia* 6, no. 1 (2018): 40–71.

37 For a company history of Jalaphrathan Cement, see: https://www.asiacement
    .co.th/en/jalaprathan-cement-plc.

38 http://www.takhli.org/rjw/tibet.htm.

39 John Dimoia, "In Pursuit of 'Peace and Construction': Hyundai Construction
    and Infrastructure in Southeast Asia," in *Engineering Asia: Technology, Colonial
    Development, and the Cold War Order*, ed. Mizuno et al. (London: Bloomsbury,
    2018): 218.

40 Dimoia, "In Pursuit of 'Peace and Construction.'"

41 Kevin Hewison, "Black Site: The Cold War and the Shaping of Thai Politics,"
    Journal of Contemporary Asia 7 (Feb 2020), https://doi.org/10.1080/00472336
    .2020.171711.

42 Porphant Ouyyanont, "The Vietnam War and Tourism in Bangkok's Devel-
    opment," *Southeast Asian Studies* 39, no. 2 (2001): 157–87.

43 Mary Beth Mills, *Thai Women in the Global Labor Force* (New Brunswick, NJ:
    Rutgers University Press, 2001), 69.

44 https://www.asiacement.co.th/en/history.

45 Johnson, *Ghosts*, 16.

46 Morita, "Infrastructuring Amphibious Space," 123

47 Danny Marks, "The Political Economy of the 2011 Floods in Bangkok:
    Creation of Uneven Vulnerabilities," *Pacific Affairs* 883 (2015): 632–51

48 Jerome Whitington, "Fingerprint, Bellwether, Model Event: Climate Change
    as Speculative Anthropology," *Anthropological Theory* 13, no. 4 (2013): 308–28.

49 Eli Elinoff, "Drawing the Future: Urban Imaginaries after the 2011 Thai
    Floods," in *The Disastrous Times: Beyond Environmental Catastrophe in Asia*, eds.
    E. Elinoff and T. Vaughan (Philadelphia: University of Pennsylvania Press, 2021).

50 Danny Marks and Louis Lebel, "Disaster Governance and the Scalar Politics
    of Incomplete Decentralization: Fragmented and Contested Responses to
    the 2011 Floods in Central Thailand," *Habitat International* 52 (2016): 57–66.

51 Danny Marks and Eli Elinoff, "Splintering Disaster: Relocating Harm and
    Remaking Nature after the 2011 Floods in Bangkok," *International Development
    and Planning Review*, 1–21; Jerome Whitington, "Bangkok's New City of Walls,"
    Anthropology News (November 15, 2019), https://www.anthropology-news.org
    /index.php/2019/11/15/bangkoks-new-city-of-walls/.

52 D'Avella, *Dreams*, 21.

53 Gabrielle Hecht, "Interscalar Vehicles for An African Anthropocene: On
    Waste, Temporality, and Violence," *Cultural Anthropology* 33, no. 1 (2018):
    109–41.

54 Paul Virilio, *Speed and Politics* (Los Angeles: Semotext(e), 2006), 29.

55 Eli Elinoff, "From Blood, Cast in Cement: Materialising the Political in
    Thailand," in *Political Theologies of Development in Asia: Transcendence, Sacrifice,
    and Aspiration*, eds. G. Bolotta, P. Fountain, and M. Feener (Manchester:
    University of Manchester Press, 2020).

4 CHEESE

xenia cherkaev
heather paxson
stefan helmreich

"Ah, Mastery of the Five Elements! ... How to tame the five essential elements of the universe – earth, air, water, fire, and cheese!"

"Cheese?"

– Rick Riordan, *The Red Pyramid*, 2010

As popular lore has it, the periodic law first came to Dmitri Ivanovich Mendeleev in a dream.[1] Having spent the night working on the second volume of his *Foundations of Chemistry*, he woke on the morning of 17 February 1869, and jotted down the first draft of his periodic system on a sheet of paper that happened to be handy.[2] The sheet, now displayed at the Mendeleev Museum apartment in St. Petersburg, was a letter from A.I. Khodnev, secretary of the Free Imperial Economic Society, and it was an invitation to Mendeleev to visit some cheese-making cooperatives near the town of Tver (see figure 4.1). Mendeleev set off for the Tver area cooperatives shortly thereafter, just as soon as he sent his one-page "Attempt at a System of Elements, Based on Their Atomic Weight and Chemical Affinity" out to the printers. He was not himself present at the March 6th meeting of the Russian Chemical Society, where his discovery of the periodic system was first announced by the Society's secretary. Although some historians have suggested that Mendeleev had simply taken ill,[3] the now commonly accepted story is that he was away on a cheese-making facilities tour.

At the same time as he dreamt of aligning the chemical elements so that their characteristics could be seen to be a periodic function of their atomic weight, Mendeleev seems also to have been preoccupied with cheese. I.S. Dmitriev, director of the Mendeleev museum in

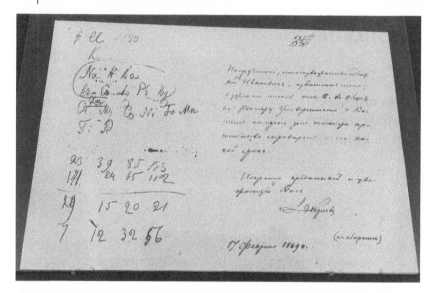

Figure 4.1. Mendeleev's first sketch for the periodic table of the elements, 17 February 1869, on view at the Mendeleev Memorial Museum Apartment in St. Petersburg (credit: Stefan Helmreich)

St. Petersburg and Mendeleev's long-time biographer, has noted that Mendeleev's "works about the Periodic law have so tightly interwoven with other, non-chemical, studies that it is often difficult to say which directions he found most important."[4] Dmitriev elaborates:

> The author of the most important discovery in history of science, being of sound health and quite cognizant of the importance of his discovery, entrusts his friend to present the first public reading of the Periodic law to a professional audience of chemistry colleagues. In the meantime, he himself makes haste to inspect the cheesemaking ventures. An unprecedented story. But, perhaps, upon returning from his trip Mendeleev did not delay to present a reading of his new discovery? He does indeed present talks on the 20th of March and the 10th of April ... but these talks are about artisanal cheesemaking and about the profitability of dairy cattle operations.[5]

What do stories of the periodic table and of cheese-making in the provinces of the Russian Empire have to do with one another? Perhaps very little. Historians have taken Mendeleev's trip as evidence of him not fully realizing the importance of his discovery.[6] But in this brief essay, we should like to assume otherwise. Holding up this historical anecdote about one of Mendeleev's side interests long enough to tease out some of its perhaps unexpected conceptual connections, we see a story not only of chemistry and cheese, but also of emergent modern social forms and philosophies. We see genres of elemental thinking from natural and social domains coming into juxtaposition with one another.[7]

Indeed, it was not the cheese itself that so directly interested Mendeleev, but the potential sale of that cheese and, more specifically, the social formations through which peasants could organize to market their dairy products in this less perishable, more profitable form.[8] Following *that* connection – among cultured dairy products, social organization, standardization, and scale – we ask what relations may be discerned between Mendeleev's interest in the cooperatives' production of standardized cheese and nineteenth-century practices of cataloguing and conceptualizing the elementary forms of social life. Mendeleev's interests are crosshatched by a classically modern tension between standardization and empirical particularity. Perhaps more surprisingly, they are also stitched together with the story of milk gone sour: a story that reaches from markets and peasants all the way down to subatomic particles.

As Mendeleev outlined in his report of 20 March 1869, the problems of peasant cheese-making were those of stabilization and scale.[9] To stabilize fresh milk into the form of cheese – and then to stabilize that cheese into the form of earned profit – peasant farmers would need to be organized at a larger scale than that of the household. Their domestic enterprises would need to be nested within larger organizational forms. For Mendeleev, writes Michael Gordin, "small-scale cheese production by independent artisans intrigued him as a possible model for organizing industry."[10] But peasant communities, emancipated from serfdom just eight years prior, in 1861, lacked institutions through which they could profitably organize their agricultural labour for the market. The most they could do was use the milk in their own households and turn it into low-grade butter and a fresh cheese made by the acidification of skimmed cow's milk, known in Russian as *tvorog*.

Neither farmhouse butter nor *tvorog* kept well, and neither were particularly profitable. Peasant dairy farmers consequently often operated at a financial loss. To help remedy this problem, the Free Imperial Economic Society (whose secretary had written to Mendeleev) had started giving out loans to finance the establishment of cheese-making cooperatives. Mendeleev's cheesemaking tour was at the behest of the Society. He was to study these ventures, of which he was already highly supportive.

The cooperatives offered peasants the ability to collectively purchase the needed equipment and hire a cheese master, and then individually contribute milk to the venture, dividing the proceeds from the sale of cheese and butter in proportion to the amount of milk contributed.[11] They were modelled upon small Swiss cheese and butter creameries, and they were also themselves a sort of experiment: part of a larger cooperative movement, first established in the mid-1860s on the model of Western European savings and loan associations. Open to members of all the estates, these cooperatives were to develop peasants into effective market producers by "merging peasant collectivism with a rationality imparted by the formally educated."[12] And, unlike the Western European associations upon which they were modelled, they did not require peasants to stake their own private property as a form of collateral. Indeed, they could not: newly emancipated peasants were legally insolvent. Peasant communities owned all major possessions collectively – real estate, equipment, seed, livestock, draft animals – making such valuable property factually inalienable for the individual peasants.[13] In this situation of non-private ownership, the cooperative was often theorized as a building block of Russian modernization.

The resemblance between the modernizing cheese factory and the periodic table of the elements that might immediately leap out is that both are organized around the assumption that complex wholes can be built up out of stable, standardized building blocks, or units.[14] That genre of conceptualizing the phenomenal world was much in elaboration in the nineteenth century, from the natural to the social sciences. Think of anthropologist Lewis Henry Morgan's claims, in his 1871 *Systems of Consanguinity and Affinity of the Human Family,* that there were basic building blocks out of which all human kinship systems were made (blood and law). Think of Herbert Spencer's 1862–93 *System of Synthetic Philosophy* and his positing of basic principles and units as animating forms of organic and social life. Darwin, Durkheim, Virchow … the list could go on. Mendeleev's work fit very much into that moment. This distinctly modern conceptualization of elements is quite different from earlier, classic notions of the "elements" as forces of nature, whether as organized into the Ancient Greek elements of fire, air, water, earth or in the classic Chinese system that names wood, fire, earth, metal, and water. The elements arranged by Mendeleev, periodically in accordance with their atomic weights, are similarly formatted building blocks of the material world rather than quasi-mystical forces of nature.

Applied to social relations, this image of stable elements allowed planners, as Anna Tsing has argued in her reflections on "scale" and "scalability," to examine the world for the small useful bits that comprise it and that may be rationally rearranged into new corporate entities.[15] This vision led to that great standardization of material and social worlds whose possibilities and perils preoccupied much of late-nineteenth and early twentieth century thought. Through the optics of scalable elements that recombined and expanded without changing their essence, everything could be seen for its relevant use: workers for labour power, mountains for ore, rivers for hydropower, persons for soldiers and citizens.

For cheese to become a "scale-making project"[16] for rural industry, it needed to be amenable to standardization and to transport. *Tvorog*, as fresh curd made for domestic consumption, was neither self-similar across farms nor sufficiently stable to travel to distant markets. Absent pasteurization and refrigerated transport, *tvorog* was not profitably scalable. Nor would it have been feasible to gather farmers' curd to salt, press and form collectively into large-format cheeses amenable to storage and travel; heterogeneous curd does not knit well together.[17] More elemental to dairying than curd, however, was the milk from which curd was made. And so, across the steppes of the Russian Empire (and, in the same decades, the Great Lakes region of the United States), farmers began pooling their fresh milk, hauling it by horse-drawn cart to centralized creameries where it would be collectively transformed into uniform curd that was drained, salted and packed into wheels for aging, transport, and storage.

Milk transformed into hard-aged cheese is more stable and standardizable than milk transformed into fresh *tvorog*. In moving dairying from peasant farm to factory cooperative, then, agricultural modernizers *changed the nature of cheese*. Peasant dairying was *not* scalable in the way Tsing describes as the ideal of modernist production, which entails expansion by adding more elements "without changing the nature of the project."[18] Indeed, the co-ops themselves were based on a Western European (Swiss) model; one co-op Mendeleev visited had even hired a Holstein couple (from the German-speaking Duchy of Holstein, which in 1866 became part of Prussia) to make cheese and butter the "Holstein-Swiss" way. Perhaps it resembled the cheese developed in the 1820s by Prussian-Swiss settlers in the eponymous East Prussian town of Tilsit. Made from skimmed cows' milk across the North Sea region in the 1800s, Tilsit (or Tilsiter) – not unlike *tvorog* – was considered "a byproduct of butter making."[19] A new "model for organizing industry" required new recipes and techniques. And, presumably, the development of new tastes and culinary repertoires.

Similarly, the cooperative form did not scale up the *tvorog*-making of peasants, unchanged. Instead, it rearranged (household) *tvorog* makers into (co-labouring) cheese makers.[20] In his 1869 speech, Mendeleev celebrates cooperatives precisely for creating a fundamentally new social unit, bound together by mutual trust and a common goal, not controlled from the top but self-governed. At their best, Mendeleev argues, these ventures should be authentically co-operational: run by peasants for their own benefit, cutting out the exploitative middleman. He was challenged, after making such arguments, about whether a cooperative of peasants alone, with no participant landowners, could generate enough milk to run a profitable cheese-making venture. But he stood firm: the most profitable ventures would be those in which peasants alone took part. Estate differences would necessarily and negatively bring with them a distrust that would threaten to sour the whole process, leading "the landowner [to think] that the peasants are tricking him, diluting their milk with water, [and] the peasants [to] think that the landowner shortchanges them."[21] The most profitable ventures would be those united by their own cheese factory, their own distribution warehouses in the urban centres, their own Russian-educated cheesemaking experts (for whose education Mendeleev asks the Society to set aside grant money). "I think," he wrote, "that the co-operative spirit employed in cheesemaking will spread to other branches of [the] peasant economy, and of course, in these other industries as in cheesemaking, great benefit may be expected of it for the producers."[22]

In the end, these cooperatives failed spectacularly: of the 14 cooperatives established by 1873 in the Tver region, "all but three were closed by 1876, and only one lasted into the 1890s."[23] The problem, as Yanni Kotsonis explains in his study of post-emancipation Russian agricultural cooperatives, was not only that this institution of ostensive peasant independence and self-reliance failed to work without non-peasant supervision and guidance, but also that there existed a fundamental tension between market and cooperative forms: "If the cooperative was to be egalitarian, it was unlikely to be economically viable [... if it] was to succeed economically, then it would benefit only a few, well-off households and by definition disrupt the image of the egalitarian commune."[24] Cooperatives, as it turned out, did not scale well: unlike replicable closed-cycle cheese factories, the cooperative form was itself rather unstable – like *tvorog*.

For Mendeleev, the substances that comprised the chemical elements were the most basic particles of matter. He rejected any notion of composite atoms as metaphysical.[25] However, by the mid-twentieth century, the elements arranged in his table were universally known by physicists to comprise not only protons, neutrons, and elections, but also quarks: odd objects that would upend any neat notion of a clean atomic ontology for the universe. Quarks cannot provide a simple, lower, more elementary level for Mendeleev's table of atomic elements, because they are unstable, erratic, and unpredictable, and known not only by their unit properties (mass, fractional charge), but by their *flavour* (though the words that describe these are not very gastronomic in character: up, down, strange, charm, bottom, and top). Moreover, the term chosen for these weird entities is itself also

suitably weird. Physicist Murray Gell-Mann, who proposed the word "quark" in 1963, drew it from James Joyce's *Finnegan's Wake*, from a polyglot passage in which the word may refer either to a volume (quart) of liquor or, more likely, a dairy product.[26] And in fact, one primary meaning of "quark" is that soft, unripened cheese that Russian peasants were making without great market success in their homes, at the time when Mendeleev inspected the cheese-making ventures: quark is *tvorog.* According to the Oxford English Dictionary, *tvorog* is in fact the etymological origin of quark.[27]

Gell-Mann was not likely aware of Mendeleev's promotion of the cooperative form, by which peasants would make durable, reproducible cheese instead of their notoriously unstable household *quark*. And Mendeleev, of course, could not have known that quarks comprised nuclear particles. But he may not have been surprised in the end. The social theories he was developing alongside discovering his periodic law rested not on self-stable units, but on shifting cooperative forms. In a pseudonymously written piece from 1877, recently translated by historian of science Michael Gordin, Mendeleev intriguingly argues for the factual impossibility of "units" in nature:

> The idea of the unit will only be true when people understand that in nature it is the same kind of meaningless thing as the zero, that the unit is nothing in and of itself, that it is only a creation of our minds similar to the ones resorted to in geometry when one imagines that a curve is composed of a large number of straight lines.[28]

Whether this view represents a change of heart for Mendeleev after long grapplings with the practical puzzles of elemental thinking (though it is probably important that he wrote this piece under a pseudonym!) is hard to know. In any event, however, *quark* as cheese and *quark* as subatomic constituent, respectively, turn out to undo the gridded promises of factory cheese production and modern chemistry both. They undo the fantasy that nested scalability is a feature of elemental nature. The periodic table was full of *tvorog* all along.

NOTES

1 P.A. Druzhinin, *The Puzzle of "Mendeleev's Table"* [Zagadka "Tablitsy Mendeleeva"] (Novoe Literaturnoe Obozrenie: Moscow, 2019), 8.

2 The dream story is commonly believed by Mendeleevologists to be apocryphal and the time within which the idea, sketch, and publication unfolded less compressed than the tale suggests (see Petr A. Druzhinin, "The First

Publication of Mendeleev's Periodic System of Elements: A New Chronology," *Historical Studies in the Natural Sciences* 50, no. 1–2 (April 2, 2020): 129–82). Note, too, that the February 17, 1869 date is in the old style Julian calendar, which lags 12 days behind the now generally accepted Gregorian. The Julian calendar was used in Russia until the 1917 October (November) Revolution.

3 Druzhinin, *The Puzzle of "Mendeleev's Table,"* 14.

4 I.S. Dmitriev, "A Desperate Soul's Protest [Dushi otchaiannoi protest]," *Vestnik Sankt-Peterburgskogo Universiteta* 4, no. 3 (2004): 115–30, 123. All translations from Russian by Xenia Cherkaev unless otherwise noted.

5 Dmitriev, "A Desperate Soul's Protest," 122.

6 Michael Gordin, "The Short Happy Life of Mendeleev's Periodic Law," in *The Periodic Table: Into the 21st Century*, ed. Dennis H. Rouvray and R. Bruce King (Baldock, Hertfordshire: Research Studies Press, Ltd, 2004), 41–90, 52.

7 On elemental thinking, see John Durham Peters, *The Marvelous Clouds: Toward a Philosophy of Elemental Media* (Chicago: University of Chicago Press, 2015); Derek P. McCormack, *Atmospheric Things: On the Allure of Elemental Envelopment* (Durham: Duke University Press, 2018); Katherine Sacco, "Elements: An #AmAnth17 Panel Review," (Member Voices, *Fieldsights*, January 18, 2018); Timothy Neale, Thao Phan, and Courtney Addison, "An Anthropogenic Table of Elements: An Introduction" (Theorizing the Contemporary, *Fieldsights*, June 27 2019); Nicole Starosielski, "The Elements of Media Studies," *Media + Environment* 1, no. 1 (2019); Dimitris Papadopoulos, Maria Puig de la Bellacasa, and Natasha Myers, eds. *Reactivating Elements: Chemistry, Ecology, Practice* (Durham: Duke University Press, 2022).

8 For a brief history of cheesemaking in the Russian empire, see Alison K. Smith, "From Gruyères to Gatchina: The Meaning of Cheese in Tsarist Russia," *Food, Culture, and Society* (forthcoming).

9 Dmitri I. Mendeleev, "About Cooperative Cheesemaking [Ob artel'nom syrovarenii]," in *Sochineniia* [Collected works], 16, 223–47 (Leningrad-Moscow: Izd. Acad. Nauk SSSR, 1951 [1869]).

10 Michael Gordin, *A Well-Ordered Thing: Dmitrii Mendeleev and the Shadow of the Periodic Table* (New York: Basic Books Gordin, 2004), 27.

11 Yanni Kotsonis, *Making Peasants Backward: Agricultural Cooperatives and the Agrarian Question in Russia, 1861–1914* (New York: St. Martin's Press, 1999), 24–5.

12 Kotsonis, *Making Peasants Backward*, 15.

13 Kotsonis, *Making Peasants Backward*, 20–4.

14 See Stefan Helmreich, "Elementary Forms of Elementary Forms, Old, New, and Wavy," in *Reactivating Elements: Chemistry, Ecology, Practice*, eds. Dimitris Papadopoulos, Maria Puig de la Bellacasa, and Natasha Myers (Durham, NC: Duke University Press, 2022), 70–83.

15 Anna Tsing, "On Nonscalability: The Living World Is Not Amenable to Precision-Nested Scales," *Common Knowledge* 18 no. 3 (2012): 505–24.

16 See Tsing, "On Nonscalability," 509.

17 In 1858, a decade earlier, in Sheboygan County, Wisconsin, a dairy farmer named John J. Smith discovered this for himself; he "bought a cheese vat and began gathering unsalted curd from neighboring farms to combine with his own. He salted, pressed, and cured the cheese, but marketing became a problem … due to its low quality and lack of uniformity. It was extremely difficult to turn out a quality cheese product when curds were collected from neighbors." This vignette is from Jerry Apps, *Cheese: The Making of a Wisconsin Tradition* (Amherst: Amherst Press, 1998), 17–18.

18 Tsing, "On Nonscalability," 507.

19 Ursula Heinzelmann, "Tilsiter," in *The Oxford Companion to Cheese*, ed. Catherine Donnelley, 712–13 (New York: Oxford University Press, 2016).

20 In the United States, this productive rearrangement was carried out through a gendered division of labour; when cheesemaking moved from farm to factory, cheesemakers shifted from farmwomen to tradesmen.

21 Mendeleev, "About Cooperative Cheesemaking," 244.

22 Mendeleev, "About Cooperative Cheesemaking," 234.

23 Kotsonis, *Making Peasants Backward*, 25.

24 Kotsonis, *Making Peasants Backward*, 26.

25 Gordin, "The Short Happy Life of Mendeleev's Periodic Law," 74.

26 In a 1977 interview with Gell-Mann, Jeremy Bernstein notes several possible sources for the passage in Joyce. Literature scholars, he writes, suggest that the reference may refer to Goethe's unfavorable comparison of cottage cheese to industrial rolled steel: "*Getretner Quark wird breit, nicht stark*" [cottage cheese that has been stepped on becomes flattened out, rather than strengthened], or that it may have been inspired by a grocery sign Joyce came across in Zurich: "Drei Mark für Muster-Quark" [three marks for model cottage cheese] (Jeremy Bernstein, "Out of My Mind: Topless in Hamburg," *The American Scholar* 50, no. 1 (1981): 7–14, 7). Quark for sale, quark squashed by the steel-rolling mills … it seems to haunt the borderlands of scalable modernization in any case. See Murray Gell-Mann, *The Quark and the Jaguar: Adventures in the Simple and the Complex* (New York: Henry Holt and Co, 1995), 180.

27 Etymology: < German *Quark*, (now regional) *Quarg* curds, cottage cheese, curd cheese (Middle High German [in late sources] *twarc*) < a West Slavonic language (perhaps Lower Sorbian *twarog*), cognate with Upper Sorbian *twaroh*, Polish *twaróg*, Czech *tvaroh*, Russian *tvorog*, of uncertain origin. With the change (within German) of initial *tw-* to *qu-* compare discussion at *qualm*.

28 Michael Gordin, "The Chemist as Philosopher: D.I. Mendeleev's 'The Unit' and 'Worldview,'" in *Mendeleev to Oganesson: A Multidisciplinary Perspective on the Periodic Table*, eds. Eric Scerri and Guillermo Restrepo, 266–78 (New York: Oxford University Press, 2018), 271.

5 COPPER

manuel tironi

"Mining runs through our veins," a Chilean mining CEO said proudly while addressing hundreds of investors in a fancy Santiago hotel.[1] By evoking the colonial-borne and state-promoted narrative of the country's intimate relation to mining, he was reassuring investors that Chile – as a history, as a culture, as a territory – was not just ready for business, but that there was no better place for it. Chile is a world-leading producer of several elemental substances, including iodine and lithium, but it is copper, the reddish metal, which is widely understood to have erected modern Chile. Chile and copper are symbiotic, or so the story goes. The country accounts for 40 per cent of the world's copper production. In 2018 alone, 5.8 billion metric tons of copper were extracted in Chile, representing 48 per cent of all exports and 9 per cent of the country's GDP.[2] So, as blood reverberates with meaning, symbolizing nationhood, place, and being, the CEO's image of mining-as-blood is aptly ontogenic: Chile as a mine and Chileans as miners. We are mining. We are from the mine. We are copper.

But the image of mining running "through our veins" also exists in another sense which, in all likelihood, the CEO would probably try to hide from investors and regulators. Mining runs in our blood, quite literally, through the toxins produced by copper's chemo-industrial processes of extraction, transportation, smelting, and refining. Copper has been metabolized, becoming a molecular element in human bodies poisoned by their exposure to the sulphur-, arsenic-, or lead-rich environments created by industrial copper extraction. In Calama, Antofagasta, Mejillones, or Puchuncaví, sites imbricated in the production chain of copper, bodies and mining, blood and copper are physiologically and territorially bound together. For if we follow the Colectivo Miradas Críticas del

Territorio desde el Feminismo[3] and "think of the body as our first territory and [recognize] the territory in our bodies," no body, including that
of the wealthy CEO, can be assessed in the restrictive grammar of the
"human." As blood cannot be disentangled from *tierra* (soil and Earth),
and as bodies always become in-place, with places always already corpore-
alized in bodies of all kinds, the violence of copper extraction on rivers,
mountains, wetlands, and valleys is also "mining running through our
veins." Using this perspective, the body becomes visible as a transactional
zone between metabolism and environment; the body becomes a knot
entangling territory, self, and affect.

If we are to piece together a new table of elements for our anthropogenic times, copper should not be addressed just in its isolated molecular
composition, as chemists might,[4] but in relation to the extended ecology
of industrial extraction. In this chapter, I turn to copper as a toxin that
pervades and harms bodies through the process of unearthing minerals
for the constitution, consolidation, and expansion of the Chilean extractivist state. I try to think about copper's chemistry through extractivist violence. But taking seriously the image of mining-as-blood, as articulated
by the CEO, I also turn to the chemistry of copper to think about the
body and its inner circulations beyond the human. If we take the body
as a category that is both compelled into the distinction between the
human and the nonhuman but also exceeds it,[5] then the chemical circulation of copper problematizes a clear-cut separation between extraction
from the (earthly) body and harm *to* the (human) body.

To delineate the sacrificial logic of copper's chemical violence – to
harm the bodies from which value is made possible – I will first introduce
the case of Puchuncaví, home of Chile's largest copper smelting plant
and the place where I have been doing ethnographic work for the last 10
years. I then take inspiration from Mapuche philosophy and Latin American communitarian feminists to propose an analytical framework that I
call "extractive hematophagy." "Hematophagy" means to feed (*phagein*)
on blood (*haima*), and by drawing attention to extractive hematophagy
I seek to reframe copper mining as an essentially vampiric process. Extractive hematophagy nurtures itself on the blood of bodies that are at
the same time human and geological, and hence reverberates with Kathryn Yusoff's concept of "white geologies,"[6] an analytic that highlights the
twinned birth of colonialism and extractivism.

Cu, atomic number 29, group 11. Copper is a native element, a category grouping nineteen chemicals known to occur as minerals. Within
native elements, copper sits with platinum, iridium, osmium, iron, zinc,
tin, gold, silver, mercury, lead, and chromium in the metallic subgroup.

While existing as a primary mineral in basaltic lavas, copper is mostly found reduced from copper compounds, either as a dissolution from magma (oxidized copper), from underwater hot springs (sulphurous copper), or from molten rocks at the interface between tectonic plates (porphyric copper). Copper, by all accounts, is coterminous with seismicity and volcanology. It was indeed in the volcanic territory of the Atacama desert where Lickanantay people in Northern Chile engaged with copper 2,500 years ago. *Payen* is "copper" in Kunza. Two and a half centuries later, copper mining is in Chile's veins and inseparable from its complex colonial history.

While a small, family-based copper mining industry flourished between the eighteenth and the nineteenth centuries in Copiapó and La Serena, copper mining as a large-scale, nation-defining operation emerged at the turn of the twentieth century. US-based companies, including the Braden Copper Company, the Guggenheim Exploration Company, and Anaconda Mining Company, controlled the industry. By the 1930s, copper extraction represented two-thirds of the Chilean economy, while 93 per cent of all copper mining in Chile was controlled by US companies. Several treaties tried to keep this unequal balance, entangling copper, Cold War geopolitics, and US neoliberal expansionism in several unsettling ways – until 1970. That year, Salvador Allende won the presidential election, becoming the first democratically elected socialist president in Chile's history. Once in office, Allende announced the nationalization of copper production, which became law one year later. "La chilenización del cobre," or the Chilenization of copper, was part of a larger program for basic resource recuperation, which included the establishment of CODELCO, the National Copper Corporation, then the largest copper company in Chile and later the largest in the world.

The "chilenización del cobre" did not end or limit the extractivist capitalist model. Rather, it enforced it, now led by and for the state. Copper was no longer in the hands of foreign capital, but was still a plentiful natural resource, "a monstrosity of nature" as a congressman of the time put it,[7] that now – more than ever – needed to be put to work for Chile's development (and revolution). Copper, as the new socialist government told it, was to be extracted and processed by brave and patriotic men to sustain Chile's "second independence."[8] Copper was not a territory, a history, nor even a chemical. Copper was, in the words of President Allende, "the wage of Chile": an economic medium of national sovereignty.[9]

After the tragic end of Allende's government, the Pinochet government's neoliberal pushback retained CODELCO while reopening mining to foreign capital, incentiventing investments with tax deductions and friendly regulations. Post-Pinochet democratic governments

consolidated and extended the neo-extractivist project incubated in the 1970s and, finally, copper mining became the most important economic sector in the country.

Along this history, copper gained agential power – not only as an economic resource but as a force shaping politics and sociality.[10] Chile's regulatory system, institutional arrangements, and territorial governance were defined by copper's forms, materialities, and placement. The country was literally infrastructured around the chemo-material specificities of copper. Today, northern and central Chile are a continuous vertebrae of mining-related operations: mine pits, housing complexes, distribution centres, tailing pools, pipelines, ore deposits, ports, freight railroads, and assorted plants – smelting plants, refining plants, carbon plants, sulfuric acid plants – unfold along the logistical requirements of the copper industrial process. Most copper mines in Chile are open pit operations, large-scale bores into mineral-rich mountains and basins with excavations of over four kilometres wide and one kilometre deep. Environmental impacts are severe. Copper ores are usually processed either with flotation or leaching techniques depending on the origin of the mineral. Both processes produce extensive chemical waste that needs to be stored, and so Chile now has 740 tailing deposits, holding billions of cubic metres of toxic residues that percolate into soils and phreatic waters. On the other end, ores are smelted in a sulfuric acid process that generates several tons per day of monoxide and dioxide sulphur gases, BTEX (benzene, toluene, ethylbenzene and xylene) compounds, lead and arsenic. The overall effect on surrounding communities' ecosystems and bodies has become a major environmental issue in Chile. The town of Puchuncaví is the most notorious example.

I meet Miguel and Rafael in Ventanas, a small fishing village within the municipality of Puchuncaví. Ventanas is home to CODELCO's copper-smelting plant, founded in 1964 with the vision of a grand copper industry for Chile that would allegedly bring riches and jobs to the region (see figure 5.1). It did not. Instead, dressed in promises, it brought along 27 other petrochemical companies and developed the dubious distinction of being the poorest municipality within the Valparaíso Region and, probably, the most polluted municipality in the entire country.

Miguel looks tired when we meet. Over the years, he has been diagnosed with several cardiovascular problems. "But I don't complain," he laughs, avoiding any self-pity. Rafael also keeps his spirits high. He has been coping with laryngeal cancer and needs to place a voice-augmenting device in his throat to make himself audible. They are both convinced that their ailments were caused or at least exacerbated by Puchuncaví's

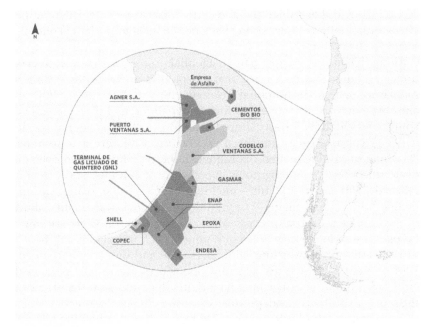

Figure 5.1.   Map of Quintero Bay and the Complejo Industrial Ventana (credit: Sebastián Saldaña)

chemical toxicity. They are longtime environmental activists, and today they want to show me one of the illegal dumping sites used by the smelting plant.

We drive to Campiche, where we cross a wire fence and climb a small hill. The soil below our feet feels extremely dry and dusty. Our boots slip and each step leaves a trail of suspended material. I can see some malvavisca (*Sphaeralcea obtusiloba*) and romerillo (*Baccharis linearis*) bushes; these are among the only plants capable of living in highly polluted soils.[11] That morning, I learn that we are walking over what ecological scientists call an "industrial barren": an extreme habitat characterized by open, bare, and severely eroded land due to the deposition of airborne pollutants, with sparse vegetation (coverage of 10 per cent or less) and acidic soils (pH < 4.0) whose reproduction capacities are severely curtailed. Industrial barrens often develop near non-ferrous smelters and refineries, primarily those of copper-, nickel-, zinc-, and lead-producing factories.[12]

When we reach the hilltop, I get the full experience of an industrial barren. This is a land whose life has been sucked up. It is not inert – as the presence of some toxic-resistant plants attests – nor dead: "dead" would be too convenient an adjective. Rather, this is a sort of zombie ecology. Life is not absent but rather pushed to its extreme, a situation of being neither fully dead nor fully alive (see figure 5.2). Looking to the smelting plant – its fumes shimmering against the backdrop of the Pacific Ocean – Miguel remembers when, back in the 1970s, cattle began to die and were found to be green inside due to copper sulphates. "And then agriculture began to die; in this area [Campiche] it died," he continues. "Nowadays it is practically a desert, a desert with some trees, with some plants and nothing else." He scrapes the dusty soil with his boot. "Nothing grows here," he adds with resignation, and his assertion stays with us as an uncanny invocation, as if summoning a presence that exceeds the abstract category of "environmental crisis" usually utilized to describe the situation in Puchuncaví. As they talk about the impossibility of a good life in Puchuncaví and harm to crops and to their own metabolisms through intimate interaction with land, water, and air, it is difficult to separate the toxicity *of* the land and *in* their bodies. Here is the tragedy of bodies, human and otherwise, worn out to maintain the continuity of copper extraction.

The toxic reverberations of copper extractivism renders visible the knots of soil and flesh in Puchuncaví, the indivisibility of ecosystems and life projects. Miguel's reference to green bodies in his account further connects land, corporalities, and suffering. *Hombres verdes* ("green men") is how neighbours in Puchuncaví describe – with both affection and disgust – former workers from the smelting plant, as they are stained with greenish wounds from copper residues in a process of accumulated sulfation.[13] For green bodies of all kinds proliferate in the toxic histories of Puchuncaví, and have come to index the creepy conditioning which the territory has been subjected to by copper extractivism. And while this conditioning messes with inside/outside, human/geological, and flesh/soil boundaries, it is also aligned along pervasive biopolitical foldings.

In Puchuncaví, sulfated copper is particularly virulent to children and women, especially pregnant and elderly women. As an effect of power geometries and patriarchal rules, the latter have no choice but to "stay with the trouble"[14] at home, at school, in the garden, and, particularly in the case of elderly women, taking care of husbands, grandchildren, crops, companion species, orchards, and cattle afflicted with toxicity. They do not just suffer the chemical inclemency of copper processing, but are also in charge – bodily, materially, affectively – of repairing the shattered basis of *buen vivir* ("the good life") in Puchuncaví. As such, they offer testimony of the harm inflicted by extractive toxins to a compounded body in which territory and metabolism cannot be disentangled.

Figure 5.2. Miguel and Rafael in Campiche (credit: Manuel Tironi)

"It's gonna be a desert here," doña Tuquita said to me once. She was born in 1931 in Los Maquis, a peasant hamlet in the Puchucaví valley, and "I haven't moved from there," she told me with pride. She has taken care of brothers, nephews, and the family land, once thriving with lentils and corn plantations and now practically dead. Doña Tuquita has acute eye problems, and as with Miguel and Rafael, she is persuaded of the liability of contamination. And, as for Miguel and Rafael, medical and ecological precarities are inseparable in her stories of decline. Trying to articulate the ambivalence between *environmental* damage and *bodily* suffering, she explained to me:

> Before, it was beautiful here, we farmed all the way to the hills. We harvested everything, we planted everything. But now it's bad. It's bad, that's what I say. Because here we had lemon trees, we had many lemon trees, a lot of lemons, and now it's scary to eat them … that's what I'm afraid of, because that [the environment] is falling down [*porque eso se está cayendo*]. Plants are been annihilated, they are drying, and all that thing [toxins] there on the soil; *all that thing that falls, falls into the ground, and into the water, and then we drink the water.*

I'm perplexed by this quasi-life, almost-death condition in Puchuncaví. A form of existence suspended just before life ends. Elsewhere I've referred to this condition as a vibration or mood,[15] trying to make sense of a toxicity that does not suffocate life completely, but weakens and abates it atmospherically, rendering meaningless the narrow category of the "environment." Copper processing dwindles life in its most ample and vital sense. Conversations with neighbours render visible this overarching exhaustion. "The area of Los Maitenes was a garden," Paulina told me once sitting in her house. "You go there now and it's depressing, it's pitiful, it's a desert ... people looking green [sick], looking tired, a lot of people are dying." In another conversation, Sara, from the small village of Pucalán, also shared her impressions and experiences living, mothering, and working in Puchuncaví, and again visions of a lurking death abound. "Imagine," she told me, "the sea doesn't have nothing anymore, it's dead, and all this is gonna be dead, from Quintero to Ventanas it's dead." These women describe the soulless perseverance of things and bodies. Life undead. A lifeless life.

By spending time in Puchuncaví I got to understand the unsettling factuality of the mining CEO's provocation. Copper actually does run through the veins of existence in Puchuncaví, as they are sacrificed to nurture and sustain copper extractivism. It is not just damage to ecosystems *and* to bodies but, as doña Tuquita suggests, it is to their emplaced entwinement. Soil and blood, water and affect, the geological and the somatic form a coherent and life-enabling yet more-than-human *body* – insofar as its circulations, transfers, and irritations cannot be taken discretely. If a body is "the material frame that gives concrete reality to a thing,"[16] then copper extraction renders evident that such framing cannot be delimited to the organized physical substance of animals and plants. Their material reality is dependent on relations with soils, waters, and atmospheres, which in turn flourish, decay, or otherwise mutate alongside anthropogenic interventions.

Simultaneously, industrial barrens, green bodies, cancers, toxic waters, and their manifold interactions make palpable copper extraction's reliance on processes that exhaust and harm this more-than-organic body. But what I learned in Puchuncaví is that this sacrificial logic does not entail destruction or complete annihilation, but rather the conditioning of a quasi-death, an almost-life, a suspended existence marked by continuous impossibility. The lives of ecosystems and humans are pushed to the brink of plausibility and are made to subsist in their barest functions. In Puchuncaví copper processing does not deny life to Miguel, Rafael, doña Tuquita, aquifers, and soils, but strips them of the possibility of a good life.

Copper, Cu, encountered as *payen* in the Andes, becomes a national commodity through the successive work of colonialism, developmentalism, and neoliberalism. From it, we can trace a trail of chemical damage in communities whose *buen vivir*, the possibility of persevering in their life projects, is sacrificed in the name of progress. The copper industry endures and is able to thrive in negativity: by harming or diminishing that from which it draws sustenance. This diminishment creates an in-between, a zone of almost-life, of life without blood, a zombie ecology. The copper industry, in short, is a vampire.

Thinking about how copper should be narrated and historized as elemental to our anthropogenic present entails foregrounding stories that are able to honour the experiences of Miguel, Rafael, and Tuquita – to attest for the afterlives of copper extractivism and the bodies with and in which copper comes to matter. How might we articulate a theory of this violence against such human-geological bodies? My inclination is to take the idea of vampirism seriously and speculate about what earlier I called extractivist hematophagy. By extractivist hematophagy I mean the vampire-like mode of existence proper of copper mining: as a state-based, large scale accumulation operation, copper extractivism is only able to endure by sucking out blood and metal from bodies – provided that these bodies are not fully human nor fully geological, but a situated material relation between both. Marx, to be sure, already warned us about the vampiric logic of capitalism. "Capital is dead labour which, vampire-like," he suggested in *Das Kapital*, "lives only by sucking living labour, and lives the more, the more labour it sucks."[17] The homology between labour and blood was quite literal. But copper extractivism adds a further layer, for what is at stake is a fundamentally different definition of what blood is and whose body this blood comes from.

Permit me the latitude to sketch, rather than detail, the theoretical genealogies inspiring the concept of extractive hematophagy. To start with, it is grounded on a specific conceptualization of bodies, blood and vampirism. Specifically, if "the racial categorization of Blackness [and Indigeneity] shares its natality with mining the New World," as Yusoff provocatively suggests,[18] then extractive hematophagy establishes a dialogue with resurgent Indigenous philosophies both as an ethical reparation against colonial (mining) violence and as an analytical gesture to foreground the *un*modern bodies at stake. Along these lines, and thinking in the context of Puchuncaví, a Mapuche settlement until the eighteenth century, the indivisibility between *mapu* ("land, place, and territory" in its ample cosmogonic sense) and *che* ("person") in Mapuche ontology provides a critical foundation to theorize bodies that are not just *in place*, but *with place* and its relations.[19] Put differently, the Mapuche body is always already a relation;[20] not the interaction of the corporeal with the geological, but the irreducible connection between them, a fundamental feature of extractive hematopaghy.

Yet, as I witnessed in Puchuncaví, extractive hematophagy is not an abstract theory of environmental harm, but a situated ethnographically inspired analytic of material effects upon specific biopolitical foldings. The notion of *cuerpo-territorio* ("body-territory") as articulated by Latin American communitarian feminists precisely tackles such entanglements, and thus offers a powerful complement to Indigenous conceptualizations. Emerging from the practices and experiences of Indigenous, Afro-descendent, and peasant women as they face the twin violence of patriarchy and extractivism,[21] the notion of body-territory highlights the ontological indivisibility of bodies and territories[22] and denounces the constant co-constitution of gendered harm against them. As stated by the Colectivo Miradas Críticas del Territorio desde el Feminismo, "We think the body as our first territory and [recognize] the territory in our bodies: when the places we inhabit are violated our bodies are affected, when our bodies are affected the places we inhabit are violated."[23] Thinking through body-territories, extractive hematophagy might be delineated as violence differentially inflicted on more-than-human bodies. As in Puchuncaví and elsewhere, extractivism and its policing affect with special intensity women, labouring men, and all those often at the frontline of their communities' pursuit of territorial justice and ethical perseverance.

To situate extractive hematophagy within the politico-territorial specificities of *mining* is equally important to emphasize its non-universality. As the case of copper in Chile shows, mining's geopolitics cannot be separated from the practices of settler colonialism, racism, and neoliberalism. Thus, Yusoff's concept of white geologies inspires extractive hematophagy, as it renders problematic the traffic between geological bodies and (Black and Indigenous) human bodies as *inhuman* matter in and for the colonial extractivist project. As Yusoff suggests, "racialization belongs to a material categorization of the division of matter (corporeal and mineralogical) into active and inert."[24] So, while Mapuche and feminist theorizations situate the body at the intersection of the corporeal and the territorial-geological as a way to indicate more-than-human relationalities and solidarities, mining makes humans and minerals functional equivalents for the operation of colonial rule and capitalist markets. Extractive hematophagy has to be theorized within this double tension: on the one hand, moving against the corporeal-geological divisibility; and on the other, enforcing exchanges between the geological and the human as fungible, extractable, sacrificial matter.

And a final thought: I can't help wondering how the chemist Dmitri Mendeleev would have reacted, 150 years ago, to the figure of extractive hematophagy, and more broadly to the treatment that Cu has received here. I want to imagine him perplexed but somehow enchanted by how copper, as a chemical element, is re-earthed into soils, bodies, and mines. His *Principles of Chemistry*,[25] the anteroom of his famous periodic table, was a conversation between chemistry and astronomy, biology, geology, and meteorology, and between basic and applied sciences; he was concerned with the coal-tar and steel industries, oil refining, and fertilizers in his country.[26] I wonder, putting extractive hematophagy in the perspective of the elemental, how much of that original hybridity was captured and retained in our current periodic table's journey from nineteenth century Russia to pollutant regulations in Puchuncaví and elsewhere. Probably none. Extractive hematophagy makes me wonder to what extent an account of the elemental attentive to its "specific social histories, material flows and inequalities," as editors of this volume have summoned, is still possible if we are attuned to Mendeleev's intuitions and adventures. Bernard le Bovier de Fontenelle, secretary of the French *Academie Royales des Science* by 1699, described the "spirit of Chemistry" as "more confused" and its principles as "more intermingled with each other."[27] So perhaps the boundary against which extractive hematophagy collides is not chemistry as a knowledge practice, but a very specific technocratic, industrialized instantiation of the elemental. Maybe extractive hematophagy, following Fontenelle's cues, is a declamation against an impossibility – the impossibility to redraw the table of elements, to infuse chemistry with critique – but also an invitation to re-engage with a more creative and open kind of chemistry.

NOTES

1   I borrow this insightful ethnographic vignette from Thea Riofrancos, "What Green Costs," *Logic Magazine*, 9 (2019, accessed 26 February 2020), https://logicmag.io/nature/what-green-costs/.

2   SERNAGEOMIN - Servicio Nacional de Geología y Minería, *Anuario de la Minería de Chile 2018* (Santiago: Servicio Nacional de Geología y Minería, 2018).

3   Colectivo Miradas Críticas del Territorio desde el Feminismo, *Mapeando el cuerpo-territorio: Guía metodológica para mujeres que defienden sus territorios* (Quito: Colectivo Miradas Críticas del Territorio desde el Feminismo, 2017), 7.

4   Michelle Murphy, "Alterlife and Decolonial Chemical Relations," *Cultural Anthropology* 32, no. 4 (2017): 494–503.

5 Following Marisol de la Cadena, "Uncommoning Nature: Stories from the Anthropo-Not-Seen," in *Anthropos and the Material*, eds. Penny Harvey et al. (Durham: Duke University Press, 2019).

6 Kathryn Yusoff, *A Billion Black Anthropocenes or None* (Minneapolis: Minnesota University Press, 2018).

7 Roberto Farias, "El Cobre Chileno Los nuevos caminos a la usurpación," *Investigación Periodística*, Serie IPE, Terram, no. 3 (2002).

8 Angela Vergara, *Copper Workers, International Business, and Domestic Politics in Cold War Chile* (University Park: Pennsylvania State University Press, 2008).

9 Vergara, *Copper Workers*.

10 Timothy Mitchell, *Carbon Democracy: Political Power in the Age of Oil* (London: Verso, 2011).

11 Rosanna Ginocchio, "Effects of a Copper Smelter on a Grassland Community in the Puchuncaví Valley, Chile," *Chemosphere* 41, no. 1–2 (2000): 15–23.

12 Mikhail Kozlov and Elena Zvereva, "Industrial Barrens: Extreme Habitats Created by Non-ferrous Metallurgy," *Reviews in Environmental Science and Bio/Technology* 6, no. 1–3 (2007): 231–59.

13 Manuel Tironi, "Hypo-interventions: Intimate Activism in Toxic Environments," *Social Studies of Science* 48, no. 3 (2008): 438–55.

14 An unexpected version of the argument in Donna Haraway, *Staying with the Trouble* (Durham: Duke University Press, 2016).

15 See Manuel Tironi, "Hacia una política atmosférica: Químicos, afectos y cuidado en Puchuncaví," *Pléyade* 14 (2014): 165–89; and Manuel Tironi, Myra J. Hird, Cristián Simonetti, Peter Forman, and Nathaniel Freiburger, "Inorganic Becomings: Situating the Anthropocene in Puchuncaví," *Environmental Humanities* 10, no. 1 (2018): 187–212.

16 Merriam-Webster, "Body," merriam-webster.com/dictionary/body.

17 Karl Marx and Frederick Engels, *Capital, Vol. 2: The Process of Circulation of Capital* (New York: International Publishers, 1967 [1885]).

18 Yusoff, *A Billion Black Anthropocenes or None*, 5.

19 Miguel Melín, Pablo Mansilla, and Manuela Royo, *Cartografía Cultural del Wallmapu: Elementos para Descolonizar el Mapa en Territorios Mapuche* (Santiago: RIL Ediciones, 2019).

20 Cristóbal Bonelli, "What Pehuenche Blood Does: Hemic Feasting, Inter-subjective Participation, and Witchcraft in Southern Chile," *Hau: Journal of Ethnographic Theory* 4, no. 1 (2014): 105–27.

21 Delmy Tania Cruz Hernández, "Una mirada muy otra a los territorios-cuerpos femeninos," *Solar* 12, no. 1 (2016): 35–46.

22 Lorena Cabnal, "Acercamiento a la construcción de la propuesta de pensami-ento epistémico de las mujeres indígenas feministas comunitarias de Abya Yala," in *Feminismos diversos: el feminismo comunitario*, ed. ACSUR (Las Segovias: ACSUR, 2010), 10–25.

23 Colectivo Miradas Críticas del Territorio desde el Feminismo, *Mapeando el cuerpo-territorio*, 7.

24 Yusoff, *A Billion Black Anthropocenes or None*, 5.

25 Bernadette Bensaude-Vincent and Isabelle Stengers, *A History of Chemistry* (Cambridge: Harvard University Press, 1996).

26 Bernadette Bensaude-Vincent, "Mendeleev's Periodic System of Chemical Elements," *The British Journal for the History of Science* 19, no. 1 (1986): 3–17.

27 Bensaude-Vincent and Stengers, *A History of Chemistry*, 42.

6 ICE

alexis rider

Easily dismissed as mere frozen water, ice is in truth far from homogenous. On land, ice forms as glaciers, as sheets, and as caps. At sea, ice floats in pancakes, floes, and bergs. Ice also cascades from the sky as hail and sleet. Accumulating, ablating, and flowing over land and sea, ice is dynamic and lively, responsive to environmental influences at a local and global scale. It is also recognized for its ability to capture natural history: ice cores are both climate proxies and containers of fossil atmosphere, preserving remnants and records of a past with which scientists model the future. But ice also articulates complex social histories. A mundane encounter with ice – a cube dropped in a drink, for example – materializes the history of the technological harnessing of cold: the collapse of space and slowing of time that helped power empire and industry since the nineteenth century.[1] A more existential encounter with ice – the immense bergs calving off the rapidly melting polar regions, for example – relays the consequences of said empire and industry: a climate crisis that collapses space and accelerates time, disproportionately impacting those least responsible for our fossil-fueled globe.

Histories, homes, landscapes, archives, climates, life: all can be refracted through ice. This ontological and epistemological flexibility is arguably reflected in ice's very materiality, which subverts classification. What *is* ice? Ice is crystal, plastic and rock; a material that can flow, blockade, and crush as effectively as it can ablate or melt.[2] As Klaus Dodds suggests, "ice is metaphorically slippery, and ice is beguiling ... It can appear, disappear and confuse those who want to identify edges, extremes and depth."[3] Spanning the spectrum from the mundane to the sublime, able to capture both natural and human history, ice is therefore an element well-positioned to articulate the multiple times and scales at work as we try to make sense of the Anthropo/Capitalo-cene.

Following the material multiplicities of ice, this essay explores several possible moments through which we can think of ice elementally. My elemental moves are limited by two broader research questions: How has ice been used as a kind of scientific instrument to understand the Earth on a planetary scale through deep time? And, how has ice shaped broader conceptions of human-geologic relations? There are myriad other ways to know ice, many with longer and richer histories than that of cryosphere research. But my focus on the scientific study of ice is intended to hew closely to the focus of this volume: the elemental, a taxonomy of nature grounded, for better or worse, in the logic of Western science. Of course, the Western perception of ice and frozen places is an extremely limited one. Many scholars have shown what ice is and means for Inuit populations in the Arctic, communities in high mountain regions, and more.[4] My efforts therefore happily fall short of anything like totality, but if elemental thinking insists on relationality, particularly relationality that slices through space and time, then incompleteness is inherent.[5] Moreover, to interrogate the Anthropocene we need to understand its ontological lineage – a lineage which is bound up in Western and scientific modes of knowledge-making.

In exploring how ice can be understood elementally, I am particularly interested in attending to questions of scale, a metric that operates on both temporal and spatial axes and brings new relations and compositions into view. To do so, I draw on Gabrielle Hecht's proposal to use "interscalar vehicles" to navigate the immense scalar varieties of the Anthropocene. In her work, Hecht follows uranium-bearing rocks across space and time, expertly holding "the planet and place on the same analytic plane," emphasizing at once the human and geologic, the political and the ethical by putting disparate registers into dialogue.[6] As I will show, ice too is an interscalar vehicle. At the scale of the cryosphere, ice is a geomorphological force, a relentless power that can literally make and move mountains. At the scale of the elemental, ice is a transmuting geochemical container, a registrar of climate change that at once holds moments of the past stable while also revealing the fluidity of all matter. These scales expose the interplay between absence and presence: a long-absent melting fossil that shaped landscape is also able to capture the past, bringing samples of a world lost to the present. And at both scales, ice has shown natural and social time to be at once entangled and co-constitutive. Expanding and contracting through time and space, ice therefore refracts the myriad dimensions of the Anthropocene.

In the following sections, I first explore how ice is a crucial part of the material substrate of the planet. As it has waxed and waned over millennia, it has helped shape the spaces and places in which life thrives.

This large-scale, geomorphological ice is, in this sense, elemental to the form of our planet. Second, I consider how scientists began reading ice as a geochemical object. Under the spectrometric gaze, ice became matter through which temporally deep planetary changes could be traced. This elemental reading rendered ice a kind of urtext for the climate crisis. And finally, I consider how, as ice has been a means of grappling with incongruous scales and temporalities, it has shaped and reshaped human-geologic relations. When we think elementally, ice – which has long been associated with lifelessness, emptiness, and stasis by the West – can be seen to be vital, teeming with time and history.[7]

As geology coalesced as a field through the early nineteenth century, meaning and order was applied to landscapes: rocks were classified, forces of erosion and accretion identified, stratigraphy codified. But some vexing geologic puzzles remained unexplainable within the framework of the nascent discipline's theory of change. One such puzzle in particular was the presence of the erratic rocks, also known as "foundlings," that dotted the British Isles. These geologic aliens bore no similarities to the landscape surrounding them. The Baron's Stone of Killochan, for example, was a thirty-seven-ton hunk of granite that rested on a Scottish hillside otherwise devoid of such material.[8] Likewise, the Converra Stone of Ireland bore no material similarity to the topography and geology of the region.[9] These immense and immovable rocks had long been explained through local myth: they had been displaced by giants, fairies, or the devil himself. Through the mid-nineteenth century, however, naturalists began to propose that an environmental agent was responsible for these foundlings: ice. While this explanation did eliminate the supernatural, it remained fantastical. To accept that a torrent of ice had moved these massive stones and then had somehow vanished entirely required a vast leap of imagination.

By the 1850s British naturalists agreed, by and large, with the claim, popularized by Louis Agassiz (1807–1873), that giant ice sheets had once subsumed large parts of the northern hemisphere and had flowed across the land crushing, dragging, and reshaping the ground beneath them as they went. Entombed in miles thick ice, foundlings were thus swept from their original homes and deposited in unfamiliar locales. As evidence mounted that "God's great plough" had shaped the surface of the Earth, familiar landscapes were transformed.[10] What was once the surface was rendered the substrate: the well-known contours of the planet's surface, the thin zone teeming with life, was in fact the negative imprint of a long-absent melted fossil. Knowledge of past ice ages therefore rendered ice agential and powerful – a force that, in tearing up the land beneath it, had made landscapes nutrient rich and amenable to life.

An element that was once constrained to liminal zones – the frozen north, mountain peaks, weeks of winter – was thus repositioned as an essential part of the deepest rhythms of Earth's history. Ice, and the ice ages, helped elucidate the scale of deep time that geologists and naturalists were grappling with. That an ice age had occurred was no longer up for debate by the 1860s, but the question of what had *caused* the ice ages was not yet answered. From 1864, Scottish autodidact and naturalist James Croll (1821–1890) dedicated his time to the question, writing in his autobiography, "Little did I suspect that it would become a path so entangled that fully twenty years would elapse before I could get out of it."[11] By connecting astronomical calculations of the Earth's eccentricity with the potential impact of feedback mechanisms (the dynamic relationship between ocean, clouds, ice, and temperature), Croll argued that ice ages were cosmologically produced, part of the universal cycles of planetary movement. Thus, with the acceptance of an icy past came a sobering prospect: over huge swaths of geologic time, what could melt could also amass. The ice sheets would, one day, return. As an interscalar vehicle, this geomorphological ice was immense both geographically and temporally, and it brought into view a secular future of certain, chilly, danger.

More than simply a large-scale natural phenomenon, the threat of a returning glacial epoch has social consequences. The spectre of returning ice sent waves through popular culture, and was the topic of articles, public lectures, and early science fiction. Journals described the ice as an ominous and lurking danger, lamenting that "the puny works of man [will be] ground and pulverized into effacement by the enormous abrading and crushing force of the moving masses of ice."[12] The fear even manifested in the popular psyche. In his *Psycho-Analytic Notes on an Autobiographical Account of a Case of Paranoia,* published in 1911, Sigmund Freud describes a man obsessed with a world-ending catastrophe, which he imagines will be caused by "a process of glaciation owing to the withdrawal of the sun."[13] Victorians thus found themselves to be "interglacial beings," existing fortuitously in a brief and unreliable moment of melt.[14]

Ice thus challenged cultural perceptions of the relationship between human and natural time.[15] In particular, it destabilized a conception of freedom, birthed during the Enlightenment, that was predicated on the separation from, and control over, nature – an ideological assertion that has undergirded histories of colonialism, slavery, resource extraction, and capitalism.[16] In Victorian Britain – a place where Anthropocentric thinking blossomed – technological changes imposed a rigid temporality on the world: standardized time guided trains, ships, factories, running the extraction of materials and labour like clockwork. The rigidity drove Charles Dickins to lament the loss of a more diurnal rhythms, writing that it was "as if the sun itself had given in."[17] Striking, then, that at such a foundational moment for the Anthropocene – just as Victorians purported to take control of nature's rhythms – ancient and absent ice undermined the claim. Deep time and vast forces of nature could not be compressed into humanity's temporal range or control. For interglacial Victorians, the deep time of ice was therefore a fundamental and humbling decentralization of the human, a stern reminder of being mired in the vastness of time and place. Ice has embodied a powerful critique of liberal anthropocentrism for centuries, insisting that there are elements that will always challenge human categorization, and shirk off human claims of control.

During a conversation with a paleoclimatologist in his office at Smithsonian National Museum of Natural History, well-insulated from the rowdy crowds that roamed the building's vast halls, I suggested that geochemistry offers up a distinct kind of time, one that pushes back against the layered, forward-marching chronos of stratigraphy. My suggestion, heavily informed by the work of Katheryn Yusoff, was met with scepticism.[18] "Well, there's only one kind of time," I was told gently. "It's more a question of resolution. Stratigraphy gives us an image of the past akin to a black and white photograph. Geochemistry gives us that same picture, but in full colour." My mind immediately jumped to a slide I had found in the Smithsonian archives, a thin-section micrograph of the Antarctic Byrd ice core, in which each ice crystal refracted light differently, producing a matrix of interlocking, vividly coloured angular shapes like a scattered handful of sea glass. The resolution certainly mattered to how scientists see ice – the vibrant micrograph contrasted dramatically with the murkily striated off-white ice cores in the other slides. But resolution isn't the end of the story. By looking "elementally," that is, by tracing the geochemistry of ice – ice as an instrument changed, and with it so did notions of temporality.

Like rock, ice can be read chronostratigraphically: the layers of matter that build up over time can be understood relationally. Ice core striations are literally counted (to a certain depth), and traces in the ice, such as ash from the eruption of Mount Tambora in 1815, used to correlate dates. That ice amassed in annual layers was recognized as early as the mid-nineteenth century. Photographs of Antarctic icebergs taken during the *Challenger* expedition of 1872–76 allowed Croll to hypothesize that the visible layers reflected annual ice buildup and could be used to calculate the age of the Antarctic ice sheet (before anyone had even set foot upon the continent). In 1930–31, Ernst Sorge (1899–1946), part of the Alfred Wegener Expedition to central Greenland, began conducting systematic pit studies that identified the accumulation of annual layers of ice. In both cases, the striated frozen layers were read as akin to stratigraphy: an accumulation that reflected linear time. But from the middle of the twentieth century other dating methods were used, such as geochemical isotope analysis, which involves examining the elemental properties of ice.[19] In geochemical analysis, the ice itself is analysed to reconstruct temperatures of the past, and the matter ice cores contain, such as air bubbles and dust particles, are extracted to generate further climatological information. Polar ice – and ice in other parts of the cryosphere – was thus recognized to hold, or in fact to be literally composed of, material essential to reconstructing and projecting past and future climates. This capacity for containment, where ice can hold time, repositioned ice. Roger Revelle (1909–1991) explained this shift to the US Congress in 1957 by describing glaciers as "a kind of library of what has happened in the past locked up and frozen."[20]

The result of this ongoing elemental analysis, which has now been applied to polar ice cores and to cores from the "third pole," high mountain glaciers, is a detailed image of much of the Quaternary Period, called "Q time."[21] In geologic terms, Q time is relatively shallow – scientists have used geochemical dating methods to understand the far deeper past of rocks. Whether applied to ice or rock, however, the techniques have revealed the "distribution and migration of the chemical elements within the earth in space and in time."[22] Seeing the "geochemical fabric of the earth," made up of ever-changing flows of energy and elements, challenges assertions about the very materiality of the world: as elements transmute, matter can change in content even if it maintains its visual form, and so too can it maintain its form even as it changes its geochemical content. The result, as Yusoff writes, is that "Anthropocene science is articulating that there is not one but many Earths, preexistent and possible, within this particular geochemical-cosmic milieu."[23] This multiplicity is revealed explicitly by ice: through isotope analysis, scientists discovered that rather than being a steady-state body that changed gradually over time, the Earth could stagger rapidly between climate states.

The elemental sense of time therefore contrasts with the chonostratigraphic time that rose to prominence as geology developed through the eighteenth and nineteenth centuries, and which underpinned early ice research. Chronostratigraphic time marches forward in a serialized manner, moving from past to future with the temporality of a Dickensian novel. Geochemical, elemental time, in contrast, is a time of starts and stops, of recombination and reconstitutions – a modernist time more akin to a T.S. Eliot poem. The temporality of ice cores therefore reveals that neither human time nor geologic time are normative, natural, or constant: both are described and materialized through specific ontological positions.

The advent of geochemistry, a "philosophical science based on the concept of the geochemical cycle in which the individual elements play their part according to established principles," reframed Earth and physical sciences far beyond ice cores.[24] More than other climate proxies of geochemical tracers, however, ice cores exemplify interscalar vehicles. Of particular note are the air bubbles in ice: small voids that are instrumental in the reconstruction of past atmospheres trapped due to the material nature of ice, which is extremely un-porous. When ice cores are extracted, the small pockets of the past – conserved in the ice under immense pressure – rapidly expand, bursting with an ear-cracking pop. This auditory articulation of the past arriving in the present captures, at a small scale, the broader temporality of the Anthropocene and the climate crisis. How the past is arriving in the present and future is complex. Carbon emissions, agricultural chemicals, nuclear fallout: all these icons of the Anthropocene have afterlives that exceed linear, stratigraphic, time.

A geochemical/elemental time of loops and flows, of dissolution and recombination, thus pushes us to rethink the Anthropocene – a term that has been counterintuitively bound to chronostratigraphy. In line with the requirements of the International Commission on Stratigraphy (ICS), efforts to identify a "golden spike" for the Anthropocene requires finding a spatially continuous mark of an abrupt temporal discontinuity in the geologic record. The radioactive isotope traces of the 1950s, when nuclear fallout was distributed around the globe, are therefore a prime candidate. Humanist critiques of origins aside, this move aligns with the intended goals of the ICS. While clearly a pragmatic means of assigning a firm start date to the Anthropocene, the need for a global marker is both totalizing and homogenizing. The most fundamental problem of the Anthropocene is that it flattens history and humanity by ignoring the racial, colonial, and patriarchal underpinnings of extraction and exploitation, and a totalized stratigraphic marker that adheres to a forward-marching, progressive sense of time seems to materialize this critique. Seeing the Anthropocene elementally, that is, as the movement of matter and energy through and across different forms (animate and inanimate, human and nonhuman, solid and ephemeral), and across deep and expansive time, can helpfully destabilize the very assumptions baked into the Anthropocene, instead revealing the multiplicity and unevenness the term must contain. It also allows us to see where the impact of humans as a geologic force has pooled, concentrated, and unevenly materialized across time and space. Thus, through geochemical time, we come to know the Earth, and our relationship to the Earth, very differently. The Anthropocene is not *just* geological, after all.

In March 2017, at the UNESCO headquarters in Paris, a new scientific initiative was announced. Titled the "Ice Memory Project" and managed by the Fondation Université Grenoble Alpes, the initiative intends to create the "first world library of archived glacier ice ... [in the] most reliable – and natural – freezer in the world": the High Antarctic Plateau.[25] By extracting ice cores from equatorial mountain glaciers – frozen places experiencing the most rapid impact of the climate crisis – and transporting them to Antarctica, the project hopes to ensure that the "scientific community of the decades and centuries to come" have access to a rapidly melting material before it is "forever lost from the surface of our planet."[26] The project is framed in urgency and the potential of future discovery. "The language of ice has an extremely broad alphabet," explains the project's website, "and we continue to discover new letters."[27] As Joanna Radin and Emma Kowal argue in *Cryopolitics: Frozen Life in a Melting World*, the storage of ice cores in laboratory freezers – artificially cold spaces that rely on energy-intensive technological temperature regulation – is "a practice saturated with ironies."[28] I would suggest that the Ice Memory Project takes these ironies further: in order to ensure future humans can understand the environmental damage wrought by fossil fuels, ice cores are relocated through energy-intensive procedures to a natural freezer that is already rapidly responding to the fossil-fuelled climate crisis.[29]

As well as illustrating the exhausting contradictions of a fossil fuelled world, the Ice Memory Project vividly underscores how ice-as-element is central to our Anthropocenic moment. If in the nineteenth century ice was repositioned as a robust geomorphological force that bound us to deep time, then in the past seventy years another repositioning has taken place. As global temperatures have risen, the element has been cast as an "endangered species," rapidly and visibly disappearing as a result of a select group of humans' excesses.[30] Melting ice will cause ocean levels to rise, rapidly redefining Earth's geography and transforming the dynamics of the Earth system. And ice will eventually self-efface, taking any records of past climates with it. Today ice, a "hot" natural solid – one that hovers close to its melting point – therefore reveals both the slow processes of the Earth and the rapid geologic impact of human-induced climate change. As glaciers retreat, sea ice thins, ice shelves crumble –

and ice cores from remote mountain peaks are squirrelled away in the Antarctic – human and geologic temporalities collapse into one another, blurring scales across space and time. The Ice Memory Project epitomizes this collapse. In order to combat the rapid heating of the Anthropocene, the sluggish deep time of the geomorphological ice sheet is mobilized to preserve the geochemical time of ice cores. Elemental kin are brought together from different parts of the Earth, to self-contain, protect, and preserve.

This temporal and spatial enfolding therefore pushes us to consider ice in its most expansive scale: as the unified, planetary amalgam called the "cryosphere." The term cryosphere was proposed by meteorologist and polar explorer Antoni Dobrowolski (1872–1954) in 1923 to describe all the frozen water on Earth, but it was taken up by the scientific community rather slowly.[31] In September 2019, however, Merriam-Webster dubbed cryosphere a "Word to Watch": as the climate crisis intensifies, the term is melting from scientific lexicon into popular discourse. The cryosphere, like the Anthropocene, is a concept that operates at a planetary register, connecting the frozen north to the frozen south, and incorporating anywhere frigid in-between. As many scholars have argued, the "planetary" as a conceptual category can be fraught. Such a universal scale can lead us to ignore the very different lived realities, and to underrepresent the deeply important – and uneven – distributions of environmental harm and responsibility. The same is true of cryosphere (or atmosphere, or lithosphere for that matter). At any planetary, spheric scale, who and what is impacted by localized change can quickly be lost.[32] But at this moment of dramatic environmental crisis, the planetary scale cannot, and should not, be ignored. As Hecht notes, if geo-engineers and ecomodernists are already happily operationalizing the planetary, environmentally minded scholars had best grapple with ways to do so, too.[33]

Interscalar vehicles enable us to encounter such scales without being lost in abstraction, bringing human and nonhuman, past and present, local and planetary into direct discourse. What socio-political histories are exposed as one sliver of the cryosphere is made to intersect with another, as high-mountain ice cores are relocated to the Southern continent? That the third pole ice was neglected as an object of concern for so long, and that an UN-adjacent agency is intervening to relocate the ice, speaks directly to the fraught history of environmentalism, so often focused on wilderness as dictated by the West. This scalar collapse likewise reveals a temporal imaginary in which the climate crisis continues unmitigated, even in the face of evidence as stark as the disappearance of an elemental planetary force. And that deep time is at once a mobilizing and paralyzing discursive tool in the wider social world is evidenced, too. This meeting of geomorphological, geochemical, and planetary ice is an interscalar story par excellence, one where the mobius strip that is at the centre of the Anthropocene – where does "human" stop and "geologic" begin? – is in full view.

NOTES

1 Rebecca H. Woods, "Nature and the Refrigerating Machine: The Politics and Production of Cold in the Nineteenth Century," in *Cryopolitics: Frozen Life in a Melting World* (Cambridge: MIT Press, 2017), 89–116.

2 The complexity of ice, in terms of its difference and content – and its status as somewhere between organic and inorganic matter – is articulated well in Julianne Yip's work on sea ice. See "Salt-Ice Worlds: An Anthropology of Sea Ice" (PhD diss., McGill University, 2019).

3 Klaus Dodds, "Geopolitics and Ice Humanities: Elemental, Metaphorical and Volumetric Reverberations," *Geopolitics* 26, no. 4 (2021): 1121–49, on 1122.

4 In particular, Indigenous communities who have lived with and on ice. See, for example, Michael Bravo and Sverker Sörlin, *Narrating the Arctic: A Cultural History of Nordic Scientific Practices* (Canton, MA: Science History Publications/USA, 2002); Julie Cruikshank, *Do Glaciers Listen? Local Knowledge, Colonial Encounters, and Social Imagination* (Vancouver: UBC Press, 2005); Karen Routledge, *Do You See Ice? Inuit and Americans at Home and Away* (Chicago: University of Chicago Press, 2018); Shari Fox Gearhead et al., eds., *The Meaning of Ice: People and Sea Ice in Three Arctic Communities* (Montreal: International Polar Institute, 2013); Kristen Hastrup, "Anticipation on Thin Ice: Diagrammatic Reasoning in the High Arctic," in *The Social Life of Climate Change Models: Anticipating Nature*, eds. Kristen Hastrup and Martin Skrydstrup (New York: Routledge, 2013).

5 Karen Barad, "After the End of the World: Entangled Nuclear Colonialisms, Matters of Force, and the Material Force of Justice," *Theory & Event* 22, no. 3 (2019): 524–50.

6 Gabrielle Hecht, "Interscalar Vehicles for an African Anthropocene: On Waste, Temporality, and Violence," *Cultural Anthropology* 33, no. 1 (February 22, 2018): 109–41, on 112.

7 Michael Bravo identifies this trend and advocates for a rethinking of the long-standing tradition in the West of seeing ice as antithetical to life. See "A Cryopolitics to Reclaim Our Frozen Material States," in *Cryopolitics: Frozen Life in a Melting World* (Cambridge: MIT Press, 2017). 27–58.

8 Archibald Geikie, "The Baron's Stone of Killochan," *MacMillan's Magazine* 17 (Feb. 1868): 312–23.

9 Robert Ball, *The Cause of an Ice Age* (New York: D. Appleton and Company, 1891).

10 Louis Agassiz famously described ice as "God's great plough." See, for example, "Ice-Period in America," *The Atlantic Monthly: A Magazine of Art, Literature, and Politics* 14, no. 81 (July 1864): 86–93, on 93.

11 James Croll, "Autobiographical Sketch," in *Autobiographical Sketch of James Croll*, ed. James Irons (London: Edward Stanford, 1896), 33.

12 "Is an Ice Age Periodic?" *Chambers's Journal of Popular Literature, Science, and Art* 10, no. 519 (Saturday, December 9, 1893): 782–4, on 782.

13 Gillian Beer recounts Freud's case in *Open Fields: Science in Cultural Encounter* (Oxford: Oxford University Press, 1996), 226.

14 Gillen D'Arcy Wood, "Interglacial Victorians," in *Victorian Sustainability in Literature and Culture*, ed. Wendy Parkins (New York: Routledge, 2018), 220–5.

15 As I argue elsewhere, British glacial theory was also used to justify assumptions about those peoples able to thrive in such cold "wasted" regions: like ice, they were positioned as dangerous threats to civilization, embedded in deep time. See Alexis Rider, "Ice Histories and Frozen Futures," in *New Earth Histories*, eds. Alison Bashford, Adam Bobbette, and Emily Kern (Chicago University Press, forthcoming).

16 Dipesh Chakrabarty, "The Climate of History: Four Theses," *Critical Inquiry* 35, no. 2 (January 2009): 197–222.

17 Quoted in Trish Ferguson, "Introduction," in *Victorian Time: Technologies, Standardizations, Catastrophes*, ed. Trish Ferguson (New York: Palgrave MacMillan, 2013).

18 Kathryn Yusoff, "Geologic Realism," *Social Text* 37, no. 1 (March 1, 2019): 1–26.

19 Chester Langway, *The History of Early Polar Ice Cores*, ERDC/CRREL TR-08-1 (Hanover, NH: Cold Regions Research and Engineering Laboratory, 2008).

20 Report on the International Geophysical Year, Hearing before the Subcommittee of the Committee on Appropriations, Eighty-Fifth Congress, First Session, 1957.

21 As of August 2019, the oldest ice core, drilled in the blue ice of the Allan Hills in Antarctica, is thought to be 1 million years old. See John A. Higgins, Andrei V. Kurbatov, Nicole E. Spaulding, Ed Brook, Douglas S. Introne, Laura M. Chimiak, Yuzhen Yan, Paul A. Mayewski, and Michael L. Bender, "Atmospheric Composition 1 Million Years Ago from Blue Ice in the Allan Hills, Antarctica," *Proceedings of the National Academy of Sciences* 112, no. 22 (June 2, 2015): 6887, https://doi.org/10.1073/pnas.1420232112.

22 Brian Mason, *Principles of Geochemistry*, 3d ed. (New York: Wiley, 1966), 4.

23 Yusoff, "Geologic Realism," 8.

24 Mason, *Principles of Geochemistry*, 5.

25 "Ice and Its Memory," Ice Memory Project, accessed August 21, 2019, https://fondation.univ-grenoble-alpes.fr/ice-and-its-memory.

26 "Ice and Its Memory."

27 "A Natural Encyclopedia of Our Environment and Climate," Ice Memory
    Project, accessed August 21, 2019, https://www.ice-memory.org/program
    /analysis/analysis-934136.kjsp?RH=1832088615292903.

28 Joanna Radin and Emma Kowal, "Introduction," in *Cryopolitics: Frozen Life
    in a Melting World*, eds. Joanna Radin and Emma Kowal (Cambridge: MIT
    Press, 2017): 3–27, on 3.

29 The project also encompasses myriad other features of Anthropocene
    science: the preservation impulse in which fragile material defends itself,
    the technological' "fix" of a "natural" freezer, and the colonial echoes of
    relocating equatorial ice, to name a few.

30 Mark Carey, "The History of Ice: How Glaciers Became an Endangered
    Species," *Environmental History* 12, no. 3 (2007): 497–527.

31 Dobrowolski wrote his book *A Natural History of Ice*, which included the
    word "cryosphere," in 1916, but it was not published until 1923.

32 That polar ice coring was given priority over high mountain glacier coring
    exemplifies this. See Mark Bowen, *Thin Ice: Unlocking the Secrets of Climate in
    the World's Highest Mountains* (New York: Henry Holt and Company, 2005).

33 Hecht, "Interscalar Vehicles," 115.

7 KEROSPHERE

zoe todd
ozayr saloojee
émélie desrochers-turgeon

The Canadian province of Alberta is a salt. It is silt. It is a solute. A precipitate. Allow us to explain. From 100 to 65 million years ago, most of what is currently known as the province of Alberta was under a prehistoric sea now known as the Western Interior Seaway. Alberta was a sea, teeming with marine life – plesiosaurs, mosasaurs, mollusks, ammonites, plankton, sharks and clams.[1] As a matter of fact, before the "big geological squeeze," before the Rockies grew and the waters receded from the prairies, moulding the recent climate, Alberta had been underwater longer than it has been "dry" land.[2]

The Western Interior Seaway stretched at times from what is currently the Arctic Ocean to the Gulf of Mexico, bisecting the current landmasses of Canada, Mexico, and the United States.[3] Each coming and going of the sea deposited on its floor a mixture of sand and silt, which slowly cemented into rock, leaving dinosaur bones in the County of Newell, coal in East Coulee, gold in McLeod River, uranium in Athabasca Lake and oil in Fort McMurray. Like a diary, these depositions contain records of time, fossil remains of critters whose life formed geological stratas into the Western Canadian Sedimentary Basin.[4]

The long process of coming out of its former marine solution and hardening into a precipitate has enriched Alberta and several other western Canadian provinces and American states with some of the most coveted concentrations of fossil fuels in the world. What happens when space and place transform from marine to prairie? What happens when the oceans retreat to leave behind landlocked cratons?[5] The traces of the ancient Western Interior Seaway enrich Alberta with a specific, complicated history that shapes every aspect of its current existence – a powerful testament to the mutability of time, place, and being.

In this chapter, we focus on Alberta in order to examine specific re-orientations of human and environmental relations in the province, the solution and dis/solution which occurred in the last 600 years because of the disruptive and totalizing forces of white supremacist colonial capitalism. We focus on Alberta as a mode of what AM Kanngieser and Zoe Todd call an environmental "kin study," rather than a case study, to better understand what geological, physical, and ideological forces shape the political economy of the territory today.[6] Alberta, as an intensely localized node of global oil and gas production, experiences complex global and local disruptions that are literally re-writing the substance of its earth, water, and atmospheres in order to serve provincial, national, and international political and economic aims.[7] Sinking into the thick elementality of the material and immaterial worlds of oil, we explore the temporalities and processes that produce Alberta's current bituminous and gaseous products and its wealths. We trouble the chemical "order of things," shift the framing of these products from commodities flowing outward to national and international markets, and posit them instead as elemental *salts* and *efflorescences* of former orders of existence being re-oriented today to fuel specific white supremacist colonial capitalist aims. In so doing, we ask what responsibilities cohere in the processes of mining, fracking, steaming, and transporting the former marine life of the Western Canadian Sedimentary Basin outwards to global markets.

With these material metaphors of salt, efflorescence, solution, precipitate, and solute we are writing our own chemistry of the "kerosphere," entangled in the petro-colonialism that currently shapes Alberta. Through this chemistry, and its de-combining and recombining relations, we urge readers to consider what it means for humans to build life and meaning on the remains of ancient seas; what it means to live in the efflorescences of the past and in the relationalities to salt as substance, metaphor,[8] precipitate and catalyst. Ultimately, we consider what it means for certain human doctrines and ideologies to force humanity to come *out of solution* – to become manifest and visible – with the substance of the world, to the point of altering its chemical makeup irrevocably. In keeping with the principle of the kin study we examine the Cree legal principle of *wahkohtowin* and the work of Leroy Little Bear to explore what legal-ethical responsibilities are required to be in better relation with the transformed phytoplankton, zooplankton and other ancient beings that are currently exploited as oil and gas in Alberta.

How do we protect the future from the present? How do we protect the past from its future? The future of Alberta's lithic and lacustrine history was partly written, first in buildings in old Montreal: in a warehouse, then a rented St. James Street home, then a neo-Renaissance building on St. Gabriel Street, complete with rusticated stone first floor and two-story pilasters. This bituminous future-history continued to be written later from Ottawa, from government buildings on Sparks Street, on Sussex Drive, from the then-named Langevin Block, where the Right Honourable Justin Trudeau now presides as prime minister. It is written now from the Office of the Prime Minister and the Privy Council, from an early twentieth century sandstone building complex, close to where we write this chapter, 3,000 kilometres away and eons apart. The Geological Survey of Canada begins in 1842 in Montreal, then settles, like a fine sediment, into the political infrastructure of Canadian government, the massif of the Department of Natural Resources.[9]

This bituminous and kerogenic[10] history of Alberta takes contemporary form in the extractive processes of both in situ and surface mining, of Steam Assisted Gravity Draining, of High Pressure Cyclic Steam Stimulation, and Submersible dewatering pumps. Draglines and bucket wheels are gone now but we have electric and hydraulic shovels and heavy haulers and a predilection for winter; for Alberta's cold, crisp air. The optimal mining temperature is -10 degrees Celsius, says Syncrude,[11] when hydrophobic bitumen is harder, when the water-coated sand is congealed and easier to drag across the earth with 4000 horsepower Caterpiller 797 Haulers. The shovel makes no distinction. Oil and ice. Bitumen, like water, can be both solid and liquid.[12] This history of mineral displacements and surfacings is written through the techno-colonial infrastructures of fluid-coking plants, and pipelines, through wellbores and boreholes, through the petro-colonial processes that violently displace time, stone, air, water and our fossil kin. These elements are now weaponized through reactors, burners, "upgradings," "refinings" and translation from avowedly heavy, dark, thick, waxy, slow matter into the "light," "sweet," and "quick"[13] synthetics that power the shovels, haulers, reactors and pumps that move this matter across land and territories, across borders and geographies, to extend kerospheric capitalism and colonialism. It is not new. *Plus ça change, plus c'est la même chose.*

We write this chapter at a distance, with the metals of the earth resonating under our fingertips, in anodized aluminium and carbon computer chassis, made with "transition metals" like nickel and vanadium found also in the bituminous sands of the Athabasca, Cold River, and Peace Lake deposits of Alberta's grounds. These are solidified metals wrenched from chemical bonds with heat and water, out from the frothy aeration of bitumen, of bitumen skimmed from its other parts and sent to refineries to be made quick, smooth, and perfectly synthetic.[14] The algae and plankton, the marine fossils of Alberta's ancient, interior

Western Seaway, are an archive of deep time and deep geology. They are an archive of deep relationality, subjected to an ontology of forceful separation and of violent transformations by technocratic machines and the geo-logics of settler colonial capitalism. The salty life of ancient seas is pulled out now from the disappeared ocean; removed from the relations formed by time, pressure, heat, and the beings who became kerogen and shale, bitumen and sand.

This wrenching out-of-solution and out-of-time and out-of-place rejects the relationality of our bodies to particles of sand, oils, airs, fossils, stones, fish, and algae. To resist this coming out-of-solution means that we must suspend the language that orients our geopolitical imaginations. It means to "write in spores," as historian Robert Macfarlane proposes in thinking about the mycorrhizal networks of the wood-wide-web, to generate a "new language altogether – one that doesn't automatically convert it [the forest] to our own values. Our present grammar militates against animacy. Our habitual metaphors subordinate and anthropomorphize the more-than-human world."[15] We write to upend and unsettle this anthropogenic and petro-colonial epistemology, to re-call an ontology of relations and relations of relations, truly seeing life in what we're calling "the kerosphere." The grammar of this lithic and lacustrine kerosphere is, as Aimé Césaire describes the power of words, "the plummet dropped in the water, the homing device that brings the self back up to the surface."[16] Attending to the kerosphere's efflorescences, we write to bring the thick viscosity of our ethical and reciprocal responsibilities to the surface.

In 1927, the Canadian chemist Karl Clark wrote that "the original source of the bitumen in the bituminous sand formation remains somewhat of a mystery."[17] In the 1920s, Clark pioneered the heat and flotation process of bitumen separation that is still used today and so, perhaps this line occurred to him in the North Lab at the University of Alberta, or at Dunvegan, where he field-tested and "ground-truthed" his ideas about this chemical (and now socio-enviro-and culturally political) process of separation. The chemistries of Clark's work ultimately rejects the pluriverse of this kerosphere, due to the pervasive slick and aggressive logics that underpin our responsibilities to these slow geologies and deep time lacustrine and lithic landscapes. We write this in a vocabulary of the kerosphere: in spores, in fossils, in trilobites and brachiopods. We write in algae and plankton, in organic compounds, in sands and drift, in Precambrian granite, and in Aptian swamps and deltas.

We write this chapter to ask for a kerospheric resistance, in particular to the white supremacist, petro-colonial orders and chemistries that weaponize kin and reciprocal relations for profit, efficiency, political ideology and ontologies of taking and extracting, of separating and de-coupling.

The "upgrading" of bitumen – the process of transforming this fossil and plant kin into oil and fuel – is also techno-scientific speech; a settler-time syntax that constructs temporal boundaries implying uselessness and out-datedness. In this insatiable quest to progress, upgrading implies better-ment, speed, linearity, the perennial battle against obsolescence, against slowness, against struggle. "Upgrade," the reactor and burner and flu-id-coker says. "Upgrade," says the hydrocracker. Upgraded now is the complex carbon, heated instead into simple chains, "upgraded" brands, weaponized kin, now in the forms of alkenes and paraffinic, naphthenic or aromatic substances. "Upgrade," begs the engine, the generator. As philosopher Karen Barad writes, the nucleus of the atom contains lega-cies of violent histories as well as the possibilities of their un-doing and alternative futures.[18] In other words, violence is written into the structure of matter through material configurations. In the case of Alberta's oil sands, the implosion of matter goes alongside the implosion of world politics – displacement, climate crisis, mass extinction, dispossession, contamination, to name a few.

Through those petro-colonial orders, complex relationality is down-graded, emptied and homogenized by these processes. Logics, orders and ontologies other than this settler colonial anthropocentric world-view are fluid-coked, hydrocracked and refined into quaint, out-of-date, out-of-time, out-of-depth platitudes. Difference is stripped of its in situ wellbores of history and relationships. They are surface-mined, crushed, and steam-pumped thousands of miles away to be co-opted by institu-tions of academe, capital, and extraction. This relationality is glossed over as ersatz pedagogies, sham political posturing, "wokeness," and spu-rious climate ethics. The anthropologist Clinton Westman calls on the importance of narrative imaginings that question Western ontologies of knowing and concepts of community when he writes:

> As components of Aboriginal knowledge, mythical understandings and ritual practices may point to some of the frailties in shortsighted, anthro-pocentric (and perhaps petro-centric) understandings of the world and of community. It is only by reflecting on such understandings, histories, and practices, on how they constitute or invert local notions of relatedness – and, correspondingly, by recognizing many state and corporate claims and proposals as potentially either monstrous or foolish – that one can come to a full understanding of the dynamics of community as currently enacted in the oil sands region.[19]

Westman's accounts of community in the oil sands region of northern Alberta exposes how bitumen becomes a multiple separation – a cleaving of relationality from ourselves and a tearing away of the non-human lives of algal and lithic kin in the kerosphere. The settler colonial project and its petrochemical purpose erase the possibilities of compassionate entanglements. In her 2018 text *A Billion Black Anthropocenes or None*, geographer Kathryn Yusoff illustrates that geology itself is racialized, thus framing extraction through a semiotic disposition that creates an "atemporal materiality dislocated from place and time."[20] The kerosphere as a conceptual tool allows for reorientation and for resituating values and discourse around kerosene. It is a non-linear framework that prioritizes generous and reparative reciprocity; a framework that emphasizes ethical imaginaries of scale (from particles to continents), of a counter-disciplinarity, of the non-human, and always within a reciprocal ecology of kindness, compassion and gentleness. The kerosphere is ordered by both deep time and also by an equally deep (fathoms deep!) continuity. The kerosphere is a new chemistry of reciprocity.

As we learn from the work of philosopher Sylvia Wynter,[21] we live in the efflorescences (precipitations) of past ideologies, the hardening of logics laid down in the silt and sand of white Western thought over the Middle Ages and Renaissance. Wynter illustrates that a dis-solution occurred during the Renaissance when white, Western Christian man was coming out of solution with its religious view of the earth-centred universe and efflorescing into heliocentrism. "Man1," as Wynter argues, was the first shift away from a specific Christian religious view of the world, and a first major step towards the horrors of white supremacist colonial imperial capitalist organization of relations across the planet.

The landscape of the Western Canadian Sedimentary Basin is its own history and its own ontology. It has written its own library and its own registers. It is already a continental tailings pond, a craton register that holds in its kerogenic, sedimentary, and efflorescing layers, the means and purpose of its own archive. These terrains are a kero-chemistry of complex connections that link the settled remains of past orders of existence – and of being – through the geologic sedimentary peel of an archaeological section, in the connection of dinosaur, plant, fish, human, trans-human, and post-human registers. Transformation is a part of life's order, of course, and radical transformations abound, unaided and un-abetted by (m)Anthropocene dictates.[22] Much of this reshaping and recasting of people and nature as units of progress in a forever scalable algorithm, however, is an arc of complicity with worldviews that, in Anna Tsing's words,

seem to be unable to do without this apparatus for making knowledge. The economic system is presented to us as a set of abstractions requiring assumptions about participants (investors, workers, raw materials) that take us right into twentieth century notions of scalability and expansion as progress ... Yet there is a rift between what the experts tell us about economic growth, on the one hand, and stories about life and livelihood, on the other. This is not helpful. It is time to reimbue our understanding of the economy with arts of noticing.[23]

Extractive colonial abstractions are coming out of solution. They have precipitated onto other orders of knowledge and being. These abstractions (no longer abstract) callously "upgrade" Alberta's physical landscapes of Taiga, Grasslands, Forest, through its muskeg and permafrost, through its freshwater and lacustrine ecosystems. These orders precipitate out onto black bear, moose, deer, bison, grasshopper, yellow-bellied sapsuckers, and fescue-grasses; spruce, pine, moss, and lichen. Almost everything in this landscape is now underscored and impacted by this petro-coloniality. It is in our bones and bloodstreams as much as it is in our orders of economies. Just as rocks are lithic tailings ponds for kerogen and bitumen, just as that kerogen was algae and plankton and trilobite and brachiopod, their "natural" transformations were through an order of a hospitable history in place. These transformations were not out of place nor written from afar. But now, all of this has been transformed into a fragmented settler colonial geological strata, weaponized as a bitter salt.

Working reciprocally in place and thinking about responsibilities towards the non-human beings who co-constitute lands, waters, and atmospheres in Alberta and in the kerosphere means that we must tend to and acknowledge the legal orders of Indigenous nations here. Legal scholar Val Napoleon[24] differentiates between state legal *systems* "in which law is managed by legal professionals in legal institutions that are separate from other social and political institutions" and Indigenous legal *orders* and Indigenous legal *traditions*. Napoleon goes on to explain that Indigenous legal orders are characterized by "law that is embedded in social, political, economic, and spiritual institutions."[25] Understanding Indigenous ontoepistemologies as *governance* systems that mediate relationships in place, and guide humans towards best relations with the more-than-human world is a critical lens through which to understand our propositions here regarding the kerosphere. This is more than simply a proposition regarding knowledge, but also about re-ordering relations in reciprocal ways to support mutual flourishing through time and space.

To reimagine human relationalities with oil and gas deposits throughout Alberta, it is vital to turn to the legal-ethical paradigms of Indigenous nations who have occupied the territories Alberta currently claims since time immemorial. While we are not able to attend in a robust way to the myriad relevant Indigenous legal orders here – be they Black-foot, Cree, Dene, Métis, Saulteaux, Stoney Nakoda, or Tsuu T'ina – we want to draw your explicit attention to non-Western onto-episte-mologies as they operate in the lands, waters, and atmospheres that comprise the Western Canadian Sedimentary Basin. This is vital as In-digenous legal orders, shaped by direct responsibilities that humans hold to non-human relations in the lands, waters, and atmospheres that comprise the prairies and boreal forest that Alberta occupies, provide deep guidance on how to be in better relation with the environment. These approaches operate in ways that challenge and disrupt colonial capitalist white supremacist logics of extraction and fragmentation, and work explicitly to centre reciprocity and relationality to humans and non-human beings.

Key to understanding Indigenous legal orders in the prairies is under-standing that, speaking very generally across contexts, humans are work-ing *in relation* to the environment as co-constituents *of* the environment, not in mastery over it. These Indigenous legal-ethical approaches refuse Western epistemologies that imagine resources as homogenous or static "commodities" (or, things) to be extracted for profit. As Vanessa Watts[26] argues, Indigenous onto-epistemologies in so-called Canada are deeply rooted in place, and centre governance built around relationships to more-than-human beings:

> habitats and ecosystems are better understood as societies from an Indige-nous point of view; meaning that they have ethical structures, inter-species treaties and agreements, and further their ability to interpret, understand and implement. Non-human beings are active members of society. Not only are they active, they also directly influence how humans organize them-selves into that society.[27]

Building on similar relatedness between Métis peoples and the more-than-human world in Northern Alberta, Métis thinker Elmer Ghost-keeper[28] delineates the difference between making a living *with* the land (in reciprocity) and making a living *off* of the land through extractive oil and gas political economies in Alberta today. By thinking carefully and ethically with the cosmological principles that Indigenous scholars working in so-called Canada, including Leroy Little Bear, Harold Car-dinal, Patti Laboucane-Benson, Vanessa Watts, Val Napoleon, Elmer Ghostkeeper and others have shared with public audiences, we can un-derstand that the environment that we are writing and thinking from in Alberta and Ontario is comprised of relations, and that these relations are co-constitutive and always shifting. Working from his perspective as a Blackfoot scholar, Leroy Little Bear explains:

> The first tenet of the Native paradigm is what we refer to as constant flux. If you were able to imagine this flux, if it was animated, you would see a constant motion, you know, of energy, waves, and so on, light and so on going back and forth. This notion of flux is a very important part of the native thinking ... things are forever always in motion, things are forever changing and so on. From that native thinking, in other words, there's nothing certain. There's nothing certain. The only thing that *is* certain is change. That is the only thing that is certain. So things are forever moving, things are forever, you know, dissolving, reforming, transforming, and so forth. That's one of the most important tenets of the Native paradigm is this notion of flux.[29]

It is difficult to be in a position of control over, let alone to extract from, onto-epistemologies that are constantly changing. Colonial capitalist markets' "dreams of alienation" require certainty, uniformity, fragmentation and order.[30] Indigenous paradigms on the prairies, broadly speaking, refuse these (il)logics of immutability. In the context of the flux that Little Bear discusses from his perspective as a Blackfoot scholar, relations are shaped by animacy. As he explains:

> A third part of the paradigm is that everything is animate. There's nothing in Blackfoot for instance, there's nothing in Blackfoot that is *inanimate* ... Everything is animate. Well, if you step back a little bit, and you say "well, if we're all animate, we all consist of those energy waves, therefore, everything – those rocks, those trees, those animals out there all have spirit, just like we do as humans." So if they all have spirit, that's what we refer to as "all my relations." You'll hear in native prayers "all my relations." When we're talking about "all my relations," we're not talking about human relations – we're talking about those rocks out there.[31]

These two principles – flux and animacy – are very important in disrupting Western petro-colonial chemistries of extraction in Alberta, and inform our understandings of how to enact reciprocity in the kerosphere. Diverse prairie Indigenous cosmologies in Alberta insistently centre principles that refuse to homogenize oil – or its prior existences as algae, zooplankton and other creatures – as *inanimate*. Instead, oil, kerogen, rocks, plants, animals and other beings are animate and in constantly shifting relation through space and time. In this way, prairie Indigenous legal orders, cosmologies and onto-epistemologies[32] operate similarly to the dynamics anthropologist Elizabeth Povinelli describes in her work on geontologies in northern Australia.[33] In order to disrupt the violence of extractive politics in Alberta, we must honour but not appropriate the Indigenous legal ethical paradigms in Alberta that refuse to entrench the binary distinctions between life and nonlife that, Povinelli argues, drives the destruction of the planet.

Nehiyaw (Cree) law centres principles of reciprocity and inter-relatedness between all things.[34] This relationality and reciprocity underpins a set of responsibilities that flow from living in coexistence with everything around us. As nehiyaw legal scholar and advocate Harold Cardinal explains, *wahkohtowin* is

> the laws governing relationships. These laws establish the principles that govern the conduct and behaviour of individuals within their family environment, within their communities, and with others outside their communities. *Wa-koo-towin* provided the framework within which the treaty relationships with the Europeans were to function. It is one of the most comprehensive doctrines of law among the Cree people and contains a whole myriad of subsets of laws defining the individual and collective relationships of Cree people.[35]

There are therefore tangible laws that govern how to be in relation with human and non-human beings in the lands, waters, and atmospheres that Alberta claims and extracts from. The principles of relationality, reciprocity, flux, and animacy all operate within prairie Indigenous cosmologies and laws, broadly, and provide guidance for how white settler paradigms must be disrupted and shifted in order to re-imagine what truly ethical, co-constitutive responsibilities to oil, kerogen, gas, and prior orders of existence can look like.

The kerosphere is a bituminous, oily, non-human ontology. The kerosphere is a stable state – or an attempt to create stabilization – in the grounds beneath our feet. It evades the greedy human desires that are meshed with violent extractive machines; with machines that bore, dig, pump, crush, heat, and transform in the mad colonial rush of continued settler capitalism. The kerosphere evades this rush, which weaponizes the ground for profit, the ground for fuel, the ground for power, the ground for powering the technologies that resuscitate and sustain extraction.

The kerosphere refutes salt as corrosion and bitumen and oil as capitalist outputs of the ground. Rather, inspired by a prairie Indigenous chemistry, an accountable Cree, Blackfoot, and Métis chemistry, this kerosphere conceptualises salts and bitumens, silts and sediments as relations. They are not violent beings but powerful structures of form and relationality that reject the notion of toxicities that Western petrochemistries posit (on repeat, on a loop, on a glitch). What is toxic, after all, can be curative in another.

The kerosphere is a co-constitution of relations that rewrites the ontologies of Western chemistries and Western science that takes, that fracks, that dominates, and that is inherent to the weaponization of fossil kin. This kerosphere is a rewriting. An algal bloom. An efflorescence. A coming out of solution.

NOTES

1   Steven M. Stanley, *Earth System History* (New York: Freeman and Company, 1999), 415.

2   Fred Bodsworth, "How the Prairies Were Made," *Maclean's Magazine*, 25 June 1955, accessed 12 December 2019, http://archive.macleans.ca/article /1955/6/25/how-the-prairies-were-made.

3   James S. Monroe and Reed Wicander, *The Changing Earth: Exploring Geology and Evolution* (Belmont, CA: Brooks/Cole, 2009), 651.

4   Steven M. Stanley, *Earth System History* (New York: Freeman and Company, 1999), 426.

5   Stable landlocked ancient continental masses which are often exposed at their surfaces.

6   A.M. Kanngieser and Zoe Todd, "From Environmental Case Study to Environmental Kin Study," *History and Theory* 59, no. 3 (September 2020): 385–93.

7   Kevin Timoney and Peter Lee, "CNRL's Persistent 2013–2014 Bitumen Releases near Cold Lake, Alberta: Facts, Unanswered Questions, and Implications: Final Report," Global Forest Watch Canada, 2014; Jeremy J. Schmidt, "Settler Geology: Earth's Deep History and the Governance of in situ Oil Spills in Alberta," *Political Geography* 78 (2020): 9.

8   Similarly to how Katherine McKittrick sat with June Jordan's kerosene and irradiation and phosphorescence, we sit with the materiality of metaphors as a creative interdisciplinary reading practice that resists the abstraction of geography. Katherine McKittrick, *Dear Science and Other Stories* (Durham: Duke University Press, 2021), 10–12.

9   The Department of Natural Resources operating under Natural Resources Canada (NRCan), is the ministry of the government of Canada responsible for natural resources, energy, minerals and metals, forests, earth sciences, mapping, and remote sensing.

10 "Kerogen is a naturally occurring mixture of organic chemical compounds that form a major chunk of organic matter in the sedimentary rocks which when heated can yield crude oil. There are two types of organic matter that are found in the sedimentary rocks – land-derived organic matter and aquatic algae–derived organic matter. When this organic matter undergoes a high amount of pressure and temperature, it gets converted into "Humin" and then into Kerogen. With time and under high pressure, the Kerogen is converted into crude oil." "What Is Kerogen?" Petropedia, accessed August 8, 2019, https://www.petropedia.com/definition/2172/kerogen.

11 "Evolution of Mining Equipment in the Oil Sand," *Oil Sands Magazine*, accessed 12 December 2019, https://www.oilsandsmagazine.com/technical /mining/surface-mining/equipment.

12 Clinton N. Westman and Katherine Wheatley, "Reclaiming Nature? Watery Transformations and Mitigation Landscape in the Oil Sands Region," in *Extracting Home in the Oil Sands: Settler Colonialism and Environmental Change in Subarctic Canada*, eds. Clinton N. Westman, Tara L. Joly, and Lena Gross (New York, London: Routledge, 2020): 160–79, DOI: https://doi .org/10.4324/9781351127462.

13 Bruce Braun, "Taking Earth Forces Seriously," in *Viscosity: Mobilizing Materialities*, eds. Karen Lutsky, Ozayr Saloojee, and Emily Eliza Scott (Minneapolis, MN: University of Minnesota Papers on Architecture, 2017).

14 See Mimi Sheller, *Aluminum Dreams: The Making of Light Modernity* (Cambridge, MA: The MIT Press, 2014), as an example of aluminium's epistemological and ontological implications and transformations.

15 Robert Macfarlane, *Underland: A Deep Time Journey* (London: Hamish Hamilton, 2019), 110–11. See also Dr. Suzanne Simard, "Mycorrhizal Networks: A Review of Their Extent, Function, and Importance," *Canadian Journal of Botany* 82, no. 8 (2004): 1140–65.

16 Aimé Césaire, in conversation with Annick Thebia Meslan, "The Liberating Power of Words: An Interview with Poet Aimé Césaire," *Journal of Pan African Studies* 2, no. 4 (June 2008): 2–11, https://www.questia.com/library/journal /1G1–192353401/the-liberating-power-of-words-an-interview-with-poet.

17 K.A. Clark and S.M. Blair, *The Bituminous Sands of Alberta* (Edmonton: W.D. Maclean, 1927), 3.

18 Karen Barad, "After the End of the World: Entangled Nuclear Colonialisms, Matters of Force, and the Material Force of Justice," *Theory and Event* 22, no. 3 (July 2019): 543.

19 Clinton Westman, "Cautionary Tales: Making and Breaking Community in the Oil Sands Region," *Canadian Journal of Sociology/Cahiers canadiens de sociologie* 38, no. 2 (2013): 228.

20 Kathryn Yusoff, *A Billion Black Anthropocenes or None* (Minneapolis: University of Minnesota Press, 2018), 4.

21 Sylvia Wynter, "Unsettling the Coloniality of Being/Power/Truth/Freedom –
    Towards the Human, After Man, Its Overrepresentation – An Argument,"
    The New Centennial Review 3, no. 3 (September 2003): 257–337. On page
    313, Wynter writes: "The paradox with which we are confronted here is
    the following: that in the wake of the intellectual revolution of the Renais-
    sance, as carried out in large part by the lay humanists of the Renaissance
    on the basis of their revalorized redescription of the human as the rational,
    political subject, Man – on the basis, as Jacob Pandian points out, of their
    parallel invention of Man's Human Others – Western intellectuals were to
    gradually emancipate knowledge of the physical cosmos from having to be
    known in the adaptive, order maintaining terms in which it had hitherto
    been known by means of the rise and development of the physical sciences.
    This meant that increasingly, and for all human groups, the physical cosmos
    could no longer come to be validly used for such projections. Instead, the
    West's new master code of rational/irrational nature was now to be mapped
    onto a projected Chain of Being of organic forms of life, organized about
    a line drawn between, on the one hand, divinely created-to-be-rational hu-
    mans, and on the other, no less divinely created-to-be-irrational animals;
    that is, on what was still adaptively known through the classical discipline of
    "natural history" as a still supernaturally determined and created "objective
    set of facts.""

22 See Kate Raworth, "Must the Anthropocene be Manthropocene?" *The
    Guardian*, 20 October 2014, https://www.theguardian.com/commentisfree
    /2014/oct/20/anthropocene-working-group-science-gender-bias.

23 Anna Lowenhaupt Tsing, *The Mushroom at the End of the World: On the Possibility
    of Life in Capitalist Ruins* (New Jersey: Princeton University Press, 2015), 132.

24 Val Napoleon, "Thinking about Indigenous Legal Orders," research paper
    for the National Centre for First Nations Governance, 2007, University of
    Toronto, Faculty of Law website, accessed 9 August 2021, https://www.law
    .utoronto.ca/sites/default/files/documents/hewitt-napoleon_on_thinking
    _about_indigenous_legal_orders.pdf.

25 Ibid, 2.

26 Vanessa Watts, "Indigenous Place-Thought and Agency amongst Humans
    and Non-humans (First Woman and Sky Woman Go on a European Tour!),"
    DIES: Decolonization, Indigeneity, Education and Society 2, no. 1 (2013): 20–34.

27 Ibid, 23.

28 Elmer Ghostkeeper, "'Spirit Gifting': The Concept of Spiritual Exchange"
    (MA thesis, Edmonton: University of Alberta, 1995).

29 Leroy Little Bear, "Native Science and Western Science: Possibilities for a
    Powerful Collaboration," filmed in Spring 2011 at Arizona State University,
    video, 1:00, accessed 12 December 2019, https://www.youtube.com
    /watch?v=ycQtQZ9y3lc.

30 See Anna Lowenhaupt Tsing, *The Mushroom at the End of the World: On the Possibility of Life in Capitalist Ruins* (New Jersey: Princeton University Press, 2015).

31 Leroy Little Bear, "Native Science and Western Science: Possibilities for a Powerful Collaboration," filmed in Spring 2011 at Arizona State University, video, 1:00, accessed 12 December 2019, https://www.youtube.com/watch?v =ycQtQZ9y3lc.

32 Vanessa Watts, "Indigenous Place-Thought & Agency amongst Humans and Non-Humans (First Women and Sky Women Go on a European World Tour!)," *Decolonization: Indigeneity, Education & Society* 2, no. 1 (2013): 20–34.

33 Elizabeth A. Povinelli, *Geontologies: A Requiem to Late Liberalism* (Durham: Duke University Press, 2016).

34 Patt Laboucane-Benson, "Historic Trauma and Supporting Aboriginal Families," *Alberta Home Visitation Network Association*, 2013, accessed 12 December 2019, http://www.ahvna.org/tiny_uploads/forms/Historic-Trauma-Article.pdf; Brenda Macdougall, *One of the Family: Metis Culture in Nineteenth-Century Northwestern Saskatchewan* (Vancouver: University of British Columbia Press, 2011).

35 Harold Cardinal, "Nation-Building as Process: Reflections of a Nihiyow [Cree]," *Canadian Review of Comparative Literature* 34, no. 1 (2007): 74–5.

8 LITHIUM

scott wark

The chemical element lithium and its molecules are the protagonists of numerous stories. In this chapter, I want to tell one: the story of lithium's discovery and ongoing use as a psychopharmacological substance. This story is perhaps not the most obvious one to tell in a book like this; today, lithium is probably best known as a key constituent of the batteries that power our mobile media devices, whose ubiquity has driven new, often violent, forms of extraction in countries like Bolivia.[1] The story I want to tell unfolds in a different register and on a different scale. It begins with bodies.

This story begins in 1949, the year an Australian clinician named John Cade published a paper outlining the antimanic effects of lithium salts.[2] With this paper, Cade announced the discovery of what many consider to be the first modern psychopharmacological substance.[3] Lithium has since become a standard psychopharmacological treatment. It is widely used as a mood stabilizer in the treatment of bipolar disorder, to treat mild depression, and in the prevention of suicide.[4] But its path to clinical acceptance was not straightforward. This is because lithium is notoriously toxic.

In this chapter, I use lithium's story as a toxic, psychopharmacological substance to rethink the relationship between chemicals and human bodies. My approach, which is derived from the history and philosophy of science – and the philosophy of chemistry in particular – begins with a somewhat paradoxical claim: what makes chemical substances *specific*, setting them apart from other kinds of substances, is their *indeterminacy*. What they *are* depends on where we find them. This approach is neither a materialist one nor a purely relational one. It is not ontological, but epistemological. To really think *with* chemicals, I argue that we have to remain responsive to their specificities as they circulate. Following lithium's circulations helps us to conceptualize what I want to call a chemical milieu: a "place" that transects bodies, chemical substances, and our ways of knowing these substances to enfold us in this – our – world. In this world, the relative, planetary scarcity of lithium helps explain why it is able to act on our bodies. It also helps us to apprehend what it is to have a body that *takes place* in chemical milieus.

Chemical substances are slippery. It is very difficult to specify something like an ontology proper to chemicals, because their possible interactions are so numerous and so environment-dependent that they are hard to determine exhaustively. For Klaus Ruthenberg, a philosopher of chemistry, this fungibility lends chemical substances what he calls an "ontological indeterminacy."[5] Unlike other substances, chemicals are defined not by their "material properties," which are subject to change, but by what Jaap van Brakel, another philosopher of chemistry, calls the "dynamic relations" they enter.[6] That is, what a chemical *is* depends on where we find it, when, and with what. This indeterminacy makes chemicals, in Ruthenberg's terms, "long-term epistemic objects."[7] They are defined by their capacity to change and, consequently, their capacity to change our knowledge of them. Chemical substances' indeterminacy makes them difficult to study, but it also makes them interesting objects to be led by.

To capture the indeterminacy that, paradoxically, makes chemicals what they are, the philosopher of chemistry Bernadette Bensaude-Vincent suggests that we study them by telling stories about them. For Bensaude-Vincent, this approach necessitates giving a "scientifically plausible" account of a chemical that explicates the "properties observed in individual substances," while also retaining a sense of their developing epistemic status.[8] That is, it uses chemicals' relationship to *time* to convey what they do as they move through different contexts. This chapter adopts this approach by drawing on a range of different literature on lithium – including from the fields of psychiatry, toxicology, and chemistry – to recover something of the ontological indeterminacy that shapes lithium's psycho-pharmacological history. As will become apparent, this approach is agnostic to ontological questions of what this chemical substance *is*, or whether it's a "material" or "real" substance. Bensaude-Vincent calls this approach "operational realism."[9] I want to treat it as an "inventive method," or a method whose tools co-constitute their object as an analysis proceeds.[10] By affirming lithium's indeterminacy and apprehending it in its relations, my gambit in this chapter is that telling lithium's story tells us just as much about the bodies that ingest it as it does about the chemical itself.

Following lithium situates the body that ingests it in what I want to call a *chemical* milieu. This concept is a more specific version of the milieu that philosophy and science and technology studies have theorized. This work is most closely associated with the vitalist philosophy of Georges Canguilhem, who noted that while living beings inhabit an exterior milieu – an environment – they're also constituted by what he calls an "interior milieu"[11]: an internal system that shapes what we are as living beings. What I'm calling the chemical milieu complexifies this – biological – concept by expanding it into a chemical register. It is tempting to conceive of chemicals through a biological lens – what they do to *this* particular body or *this* particular environment – or to reduce them to physical causes – like material properties.[12] But philosophers of chemistry argue that either manoeuvre is reductive. As Gaston Bachelard notes, chemical substances do not have the same degree of "coherence" at all levels of existence.[13] Chemicals' ontological indeterminacy *scales*, cutting across science's disciplinary demarcations – and distinctions like body and environment, or interior and exterior milieus – as they circulate through different contexts.[14] To understand a chemical on its terms, we have to follow its circulations in its own milieu, which enrols bodies and their interiors, planets and their deep histories, into a "place" that allows us to see each of them differently.

The body that ingests lithium doesn't change – grasping it doesn't require a wholly new conception of what is. Rather, we could say that it *takes place*, differently. What changes is how it might be known and, by extension, how we might use it to know *our* place in *this* world, as we're constituted by chemicals and their circulations. To discover how, we have to follow lithium's circulations. This particular story starts in a pair of unlikely places: a prisoner of war camp in Singapore and an asylum in a town called Bundoora, 16 kilometres north of Melbourne, Australia.[15]

During the Second World War, John Cade had served in the Australian Army as an officer and physician. In 1942, he'd been posted to the Pacific and stationed in Singapore, where he was unfortunate enough to be captured when the city fell to Japanese forces. Cade spent the rest of the war interned in the notorious Changi prisoner of war camp, where he had continued his work as a physician, treating Australian prisoners for ailments varying from malnutrition to mental disorders. As part of this work, Cade continued to conduct standard post-mortem examinations on patients who died in his care. Those patients who had had a mental disorder all seemed to have a significant pathology, like a tumour, in their bodies. This observation led Cade to hypothesize – erroneously, though in accordance with other clinicians of the period[16] – that mental disorders might have a physical cause. This would prove to be the germ of his discovery of lithium's psychopharmacological effects.

Cade brought this hypothesis with him to Bundoora and tested and refined it while ministering to his patients. He began to wonder whether "manic-depressive insanity," as bipolar disorder was then known, might be caused by a toxic substance whose cyclic excess or lack corresponded to sufferers' respective high and low moods.[17] Cade surmised that such a toxic substance must be eliminated in the waste products of a patient's body, so he tried to isolate it in urine samples taken from patients with mania, schizophrenia, and depression. He injected these samples into the stomach cavities of guinea pigs to see if some were more toxic than others; curiously, samples from his bipolar patients were. Encouraged, Cade went on to isolate urea as the toxic agent in more experiments with guinea pigs. However, his plans to conduct further experiments with urea were hampered by its low water-solubility, which made it hard to inject into his guinea pig subjects. By happenstance, he got around this problem by choosing the most soluble salt formed by uric acid – lithium urate – to formulate a new solution. Stunningly, this solution turned out to be less toxic.

Cade surmised that the lithium must have some kind of mitigating effect on urea's toxicity. To isolate the lithium's effects from urea's, Cade added lithium carbonate to his lithium urate solution; later, he experimented with a solution of lithium carbonate on its own. The results were the same. As Cade described it in 1949, "after a latent period of about two hours the animals, although fully conscious, became extremely lethargic and unresponsive to stimuli for one to two hours before again becoming normally active and timid."[18] This sedative effect interested Cade more than his original hypothesis about toxic substances causing mania and depression. He formulated a new hypothesis: that lithium carbonate might have a sedative effect on patients with mania.

Cade had little extant data about lithium to work with. Though the substance had a pharmacological history – it was used to treat gout more than half a century earlier[19] – data on dosages were inconclusive. Cade decided to test the substance on himself first and, after deciding that its potential therapeutic effects outweighed its potential toxicity, began the trials that would inform his 1949 paper.[20] Lithium's effect on Cade's manic patients was "dramatic": the case histories of the 10 patients he treated at Bundoora detailed in his paper reports that it mitigated his patients' manic symptoms.[21] But so too were its potentially fatal side effects. Cade noted many symptoms that are still warned against today, observing that an overdose made patients look "ill – pinched, drawn, grey and cold," going on to say, "Unless such symptoms are followed by immediate cessation of intake there is little doubt that they can progress to a fatal issue."[22] This fate would ultimately befall the most successful case in his original paper, a patient identified only as W.B. This patient was originally discharged from Bundoora after a course of lithium treatment, only to relapse and return. Soon after being put back on lithium, he would die from a combination of lithium toxicity and an infection, causing Cade to abandon lithium for several years after this patient's death.[23]

The year 1949 couldn't have been a worse time for Cade to publish his findings. In the United States, lithium salts had been prescribed to patients with cardiac and renal conditions as a substitute for table salt until 1949, when they were banned for causing several deaths.[24] Lithium's psychopharmacological fortunes wouldn't turn until 1954, when a Danish psychiatrist called Morgens Schou published the results of a double-blind trial – in another milestone, the first of its kind in psychiatric medicine – demonstrating its efficacy and establishing safe dosages.[25] Despite this and later evidence, the U.S. Food and Drug Administration would not lift its ban until 1970.[26] Though lithium is widely accepted today, its toxicity still impacts how it is perceived. In a recent large-scale meta-analysis of its use and effects, McKnight et al. conclude that while its clinical uses are proven, it continues to be "hampered" by what they coyly describe as "safety concerns."[27]

Cade's discovery of lithium's antimanic properties traced a path from guinea pigs to patients via his own body. With the administration of lithium carbonate, each of these bodies became an experimental site. They helped to produce an ambivalent knowledge: of a psychopharmacological substance that could treat a condition, mania, for which there had been no extant treatment, and of a substance notorious for its ill-effects, including toxicity and death. This ambivalent knowledge has been lithium's leitmotif ever since.

This ambivalent quality suggests that we might conceptualize lithium as a *pharmakon*: a substance that can either harm or heal, depending on its dose.[28] The *pharmakon*'s ambivalence is typically used to undermine the coherence and normativity of the thing in question. A medicine might be good or bad for a patient – and their body – depending on its dose. But I want to resist reading lithium in this way. The medical literature conceives of a chemical's passage through the body using two different terms: pharmacodynamics, or what a chemical does to the body; and pharmacokinetics, what the body does to that chemical.[29] This division reminds us that chemicals' circulations are relational and that chemicals can be the *pharmacokinetic* subjects of their own story.

The question I want to ask is: What does lithium's toxic ambivalence tell us about how we come to know bodies, as our bodies and as bodies that take place in *a* – chemical – world? Like Cade's guinea pigs, his body, and the bodies of his patients, the bodies lithium circulates in today remain experimental sites. To understand why, we need to return to lithium's circulations once again, with more specificity.

This story starts in the mouth. Lithium is ingested in the form of a soluble salt – either a lithium carbonate tablet or a lithium citrate solution – which is taken two or three times a day in a dose ranging between 600 and 1800 milligrams.[30] From here, lithium passes into the gut, where it is quickly absorbed into the gastrointestinal tract, enters circulation in the blood and, from there, enters the interstitial fluids found around the body's cells. Over a period of 6–10 days, lithium will finally pass in to these cells by taking the place of other salts, like potassium or sodium, in transport proteins that aid the movement of small molecules across cellular barriers.[31] It is here that it acts on the patient who ingested it. Lithium does not stay in the body for very long. It has a half-life averaging twenty to twenty-four hours – often longer in older patients – and is eliminated almost entirely by the kidneys.[32]

This bare description tells us little about what lithium is. I introduce it not to try to explain lithium's psychopharmacological mechanisms – in fact, clinicians and researchers do not actually understand the mechanisms behind psychopharmacological effects which are, as Ross J. Baldessarini puts it, "complex and elude a simple, coherent, and comprehensive theory of its stabilizing effects on mood and behavior."[33] I introduce it to begin to articulate how lithium draws the bodies that ingest it into a chemical milieu.

Because of its toxicity, patients who take lithium have to be particularly careful about the amount of it in their body at any given time. They have to monitor its adverse effects on their body, which scale in severity. These effects are gradated, falling into common, less common, and rare – or toxic – effects. Common effects include confusion, stomach upsets, dry mouth, headache, nausea, thirst, increased urination, weight gain, and fine tremors. Less common effects include goiter, hypothyroidism, and motor problems. Its rare, toxic effects include ataxia, hyperthyroidism, seizures, renal failure, and, if concentrations remain high, death.[34] Crucially, these effects can interact and compound the likelihood of toxicity. For instance, lithium is administered as a salt. Its concentration is affected by the amount of sodium and water in the body. Too little sodium and lithium will substitute for it, sending its concentrations spiking. Yet lithium also causes increased urination. It can cause dehydration and lowered electrolyte levels, precipitating its substitution for sodium and, again, increasing its concentration.[35] Like the antidepressants Elizabeth

Wilson discusses,[36] lithium engages multiple bodily systems – ingestion, digestion, perspiration, circulation, cellular functioning, and elimination – and basic bodily practices – eating, moving, and sleeping. Its passage isn't linear; it *circulates*, subjecting the body to peaks and troughs and to interlinked effects. It turns the body into a specific kind of long-term epistemic object.

Patients treated with lithium have to be able to register whether they have a toxic concentration in their body. They might know this by noting, for instance, that motor problems are a sign that they could be developing a possibly fatal concentration. In other words, an adverse effect might be an index of this indeterminate, circulating substance's status in their body. But they will also know this status by another kind of index: a number. Patients on lithium are likely to be subjected to monitoring regimes that measure the level of lithium in samples of their blood serum, expressed as millimoles per litre (mmol/L) measurements. Along with kidney, thyroid, and electrolyte tests, these are taken every 1–2 weeks when titrating and every 3–6 months when a maintenance dose is achieved. This measurement is compared to what is known as lithium's "therapeutic index," or a ratio that establishes the threshold beyond which mmol/L concentrations of lithium are likely to be toxic. This index allows us to quantify lithium's toxicity: it has a very narrow therapeutic ratio of 2:1, which means that twice the therapeutic concentration in the body risks rare adverse effects – including death.[37] In practice, the therapeutic index is a heuristic that guides the treatment of patients by establishing a window for efficacious therapeutic concentrations. I want to suggest that the therapeutic index contains within it an entire theory of the relationship between chemical substances and the body that is put in to practice every time it is used in a clinical setting.

The therapeutic index can be understood as a medicalized application of what Bachelard called "phenomenotechniques."[38] Bachelard argues that certain classes of scientific phenomena – chemicals among them – can only be known once they are "selected, filtered, purified, shaped by instruments."[39] This portmanteau denotes an epistemology that is produced when technology becomes necessary to render phenomena knowable. Bachelard also conceives of these phenomena as "substantialized substantives," or substances that only are insofar as their qualities are "regarded as a system" and "unified in a role."[40] The therapeutic index defines lithium's role in the negative: one must not take so much that it becomes toxic. At the same time, it enrols lithium into a specific system: the body, the blood sample, the phlebotomist, the test, the result. This system makes toxicity *knowable* – to a degree.

A patient on lithium might *know* that their blood serum levels have to fall within a set of ranges: for depression-centric bipolar disorder or when used as a complement to an anti-depressant, 0.4–0.8 mmol/L; for acute mania, 0.6–1.0 mmol/L. At concentrations below 0.4/0.6 mmol/L, lithium will not work as a prophylaxis against mania. At concentrations above ~1.3 mmol/L, patients risk "acute" toxicity.[41] But these concentrations are never stable. They fluctuate over time, reaching their highest levels in the few hours after lithium is ingested and distributed through the body, before tapering to a stable concentration in the hours that follow.[42] Phlebotomists typically test what's called a "trough" measurement, which is measured in samples extracted when fluctuations settle to their daily nadir around 12 hours after a dose is taken.[43] In circulation in the body, lithium institutes toxicity as a constant, fluctuating mode of embodiment. This mode of embodiment is transected by multiple temporalities: the dose; the body's reaction 2, 12, and 24 hours later; the weekly, fortnightly, or three-monthly tests; the asymptotic horizon, treatment's cessation. These are the tempos of a body transected by the chemical milieu.

What is known about this body, by the patient who ingests lithium or by the clinician who treats them, is doubly indexed. The patient indexes the chemical's impact on their body by its adverse effects, registered as symptoms, which are noted down or realized in dialogue with their clinicians. Then, these effects must be indexed again to renal and thyroid function, electrolyte levels, and, crucially, blood serum concentrations. Bachelard argues that the instruments used to produce phenomenotechniques can be thought of as "theory materialized," imprinting the substances they produce with their constitutive assumptions.[44] Chemical, for the patient treated by lithium, becomes *measure*: of one's wellbeing; of one's tolerance for side effects; of one's quantifiable proximity to mortality at a given moment in time. The tests the lithium patient is subjected to substantialize blood differently, underscoring both the blood's chemical nature and its ontological indeterminacy. By isolating the blood as an epistemic object, these tests – and the therapeutic index – exteriorize what Canguilhem called an "interior" milieu. This exteriorization situates the body in a chemical milieu.

Through the medium of specific phenomenotechniques, lithium's circulating presence transects inner body and outer world. This body is orientated around a specific kind of ontological indeterminacy, informed by peaks and troughs, concentrations and thresholds, adverse and beneficial effects. This indeterminacy is partial, and yet not. It is partial, because it effects only some of the body's systems. It is not partial, because it can occasion death – and a body is never more a whole than when it is no longer living. In their indeterminacy, chemicals situate us in a world. This is the chemical milieu, a world occasioned by a chemical relation. To end, I want to make a speculative proposition about how this world might scale – and why it cannot be generalized. The indeterminacy of chemicals necessitates that we emphasize their specificities when we tell their stories, because the chemical milieus they open up are specific: to us, to the world that takes place with their circulations – and to *our* world.

This story is our world's story. In our world, lithium's circulations have certain specificities. Though we might not know exactly how its psychopharmacological mechanisms work, some scientists have nevertheless speculated about why it is able to affect us the way it does. In an essay on lithium, the eminent late inorganic chemist R.J.P. Williams notes that "life, as we know it, is confined to the chemistry of certain elements" – its building blocks.[45] Within this range, though, Williams argues that "there is no reason, *apart from the availability*, why biology should not have built bones from strontium, barium, or beryllium salts or why it should not have learned to use the movement of *lithium*, rubidium, or caesium ions to govern nerve impulses."[46] Our bodies do not use lithium or these other elements, except in trace amounts, because of their relative scarcity on Earth. Lithium is an unstable element. Moreover, the distribution of elements is "frozen in planetary systems" when they are formed.[47] Taken speculatively, this proposition *scales* lithium's chemical milieu. It places us in, and on, a world: *ours*. This world emerges at the point of contact between lithium and body, as a micro-scale presence whose effects emerge precisely because of a macro-scale *absence*.

So, we exit the body and find ourselves among the elements. I'd like to suggest that this story has two consequences: one methodological, one conceptual.

Bodies, as we are no doubt used to hearing, are multiple. The body undergoing lithium treatment is multiple in a specific way. In their ontological indeterminacy, chemicals enjoin us to refrain from positing ontological propositions in advance of following their circulations. If lithium is able to act on our bodies because of its planetary-scale absence, it is not compatible with many of the prevailing materialisms that inflect our concepts of chemicals, because we cannot say in advance what its matter *is*. We can only know it, in an epistemological register, by what it becomes. This is the conceptual consequence.

Chemicals enjoin us to adopt what François Dagognet calls a "regional epistemology," or an epistemology that acknowledges that "the human body already contains several "bodies" within it, and they are reflected in one another."[48] The body in which lithium circulates, in the chemical milieu it unfolds, is an interiority exteriorized in a particular way: through phenomenotechniques that index toxicity and that allow patients to mediate their bodies using a metric. This body, and its mediated part-whole relations, is a chemical one; the other relations in which we're entangled articulate others, which may or may not intersect. This is the methodological consequence, whose starting point is the recognition that *we don't yet know what a chemical can do*. Chemicals nest indeterminacy: theirs, their milieu's, our body's, our world's. The only way to think with these long-term epistemological objects is to join their circulations and to find what they enfold.

NOTES

1   Anna C. Revette, "This Time It's Different: Lithium Extraction, Cultural Politics and Development in Bolivia," *Third World Quarterly* 38, no. 1 (2017): 149–68.
2   John F. Cade, "Lithium Salts in the Treatment of Psychotic Excitement," *Australian and New Zealand Journal of Psychiatry* 16 (1982): 129–33.
3   Frederick K. Goodwin and S. Nassir Ghaemi, "The Impact of the Discovery of Lithium on Psychiatric Thought and Practice in the USA and Europe," *Australian and New Zealand Journal of Psychiatry* 33 (1999): S54–64; Ross J. Baldessarini, *Chemotherapy in Psychiatry: Pharmacological Basis of Treatments for Major Mental Illness* (Dordrecht: Springer, 2013); David Healy, *The Antidepressant Era* (Cambridge, MA: Harvard University Press, 1997).
4   Ramadhan Oruch, Mahmoud A. Elderbi, Hassan A. Khattab, Ian F Pryme, and Anders Lund, "Lithium: A Review of Pharmacology, Clinical Uses, and Toxicity," *European Journal of Pharmacology* 740 (2014): 464–73; Rebecca F. McKnight, Marc Adida, Katie Budge, Sarah Stockton, Guy M. Goodwin, and John R. Geddes, "Lithium Toxicity Profile: A Systematic Review and Meta-Analysis," *The Lancet* 379, no. 9817 (2012): 721–28; Gin S. Malhi and Samuel Gershon, "Ion Men and Their Mettle," *Australian and New Zealand Journal of Psychiatry* 43 (2009): 1091–95.
5   Klaus Ruthenberg, "About the Futile Dream of an Entirely Riskless and Fully Effective Remedy: Thalidomide," *HYLE–International Journal for Philosophy of Chemistry* 22 (2016): 56.
6   Jaap van Brakel, "Philosophy of Science and Philosophy of Chemistry," *HYLE: International Journal of Philosophy of Chemistry* 20 (2014): 24.
7   Ruthenberg, "About the Futile Dream," 56.
8   Bernadette Bensaude-Vincent, "Philosophy of Chemistry," in *French Studies in the Philosophy of Science: Contemporary Research in France*, eds. Anastasios Brenner and Jean Gayon (Dordrecht: Springer, 2009), 169.
9   Bensaude-Vincent, "Philosophy of Chemistry," 180.
10 Celia Lury and Nina Wakeford, *Inventive Methods: The Happening of the Social* (London and New York: Routledge, 2012), 12–13.

11 Georges Canguilhem, "The Living and Its Milieu," in *Knowledge of Life*, eds. Paola Marrati and Todd Meyers (Fordham University Press, 2008), 111.

12 Łukasz Lamża, "How Much History Can Chemistry Take?," *HYLE – International Journal for Philosophy of Chemistry* 16, no. 2 (2010): 117.

13 Gaston Bachelard, *The Philosophy of No: A Philosophy of the New Scientific Mind*, trans. G.C. Waterston (New York: Orion Press, 1968), 46.

14 Roald Hoffmann, "What Might Philosophy of Science Look Like If Chemists Built It?," *Synthese* 155 (2007): 334; Bensaude-Vincent, "Philosophy of Chemistry," 169.

15 Unless noted, Cade's biographical details are sourced from Greg de Moore and Ann Westmore, *Finding Sanity: John Cade, Lithium, and the Taming of Bipolar Disorder* (Sydney: Allen & Unwin, 2016).

16 David Healy, *The Creation of Psychopharmacology* (Cambridge, MA, and London: Harvard University Press, 2002).

17 Philip B. Mitchell and Dusan Hadzi-Pavlovic, "Lithium Treatment for Bipolar Disorder," *Bulletin of the World Health Organization* 78, no. 4 (2000): 515.

18 Cade, "Lithium Salts," 130.

19 Psychiatric researchers disagree whether Cade could be said to have "discovered" the substance. Shorter and Malhi and Gershon note that Carl and Frederik Lange used lithium to treat depression in 1894, far predating Cade. Others, including Baldessarini and Mitchell and Hadzi-Pavlovic, argue that the Lange brothers' discovery was lost to medical science and that, while Cade was aware of lithium's use for gout and even of the mood-altering properties ascribed to lithium-rich natural springs, he was not aware of the Lange's use. I adopt the latter interpretation. See Edward Shorter, "The History of Lithium Therapy," *Bipolar Disorders* 11, no. s2 (2009): 4–9; Malhi and Gershon, "Ion Men," 1094; Baldessarini, *Chemotherapy in Psychiatry*, 92; and Mitchell and Hadzi-Pavlovic, "Lithium Treatment," 516.

20 Goodwin and Ghaemi, "The Impact of the Discovery of Lithium," S58.

21 Mitchell and Hadzi-Pavlovic, "Lithium Treatment," 516.

22 Cade, "Lithium Salts," 133.

23 Goodwin and Ghaemi, "The Impact of the Discovery of Lithium," S58.

24 Baldessarini, *Chemotherapy in Psychiatry*, 94; Shorter, "The History of Lithium Therapy," 2.

25 Goodwin and Ghaemi, "The Impact of the Discovery of Lithium," S59.

26 Shorter, "The History of Lithium Therapy," 4.

27 McKnight et al., "Lithium Toxicity Profile," 721.

28 Jacques Derrida, *Dissemination*, trans. Barbara Johnson (London: The Althone Press, 1981); Elizabeth A. Wilson, *Gut Feminism* (Durham and London: Duke University Press, 2015).

29 Wilson, *Gut Feminism*, 99.

30 Baldessarini, *Chemotherapy in Psychiatry*, 94; Shorter, "The History of Lithium Therapy," 95–6.

31 Alfred Bernard, "Lithium," in *Handbook on the Toxicology of Metals*, 4th ed., eds. Gunnar F. Nordberg, Bruce A. Fowler, and Monica Nordberg (Cambridge, MA: Academic Press, 2014), 971; Oruch et al., "Lithium: A Review of Pharmacology," 467.

32 Researchers give different figures for lithium's half-life. I have used Baldessarini's because it represented the median. Aral and Vecchio-Sadus give it as 24 hours; Bernard as 18 hours in young patients and 36 hours in older ones, respectively. The assessment by Oruch et al. of 12 hours seems an outlier and has been disregarded (2014: 467). See *Chemotherapy in Psychiatry*, 98. Hal Aral and Angelica Vecchio-Sadus, "Toxicity of Lithium to Humans and the Environment – A Literature Review," *Ecotoxicology and Environmental Safety* 70, no. 3 (2008): 350; Bernard, "Lithium," 971; Oruch et al., "Lithium: A Review of Pharmacology," 467.

33 Baldessarini, *Chemotherapy in Psychiatry*, 94; see also Bernard, "Lithium," 972.

34 Duarte Mota de Freitas, Brian D Leverson, and Jesse L Goossens. "Lithium in Medicine: Mechanisms of Action," in *The Alkali Metal Ions: Their Role for Life*, eds. Astrid Sigil, Helmut Sigil, and Roland K.O. Sigel (Zurich: Springer, 2016); Baldessarini, *Chemotherapy in Psychiatry*; Mitchell and Hadzi-Pavlovic, "Lithium Treatment"; Oruch et al., "Lithium: A Review of Pharmacology."

35 Aral and Vecchio-Sadus, "Toxicity of Lithium," 351; Bernard, "Lithium," 972.

36 Wilson, *Gut Feminism.*

37 Oruch et al., "Lithium: A Review of Pharmacology," 467.

38 Gaston Bachelard, *The New Scientific Spirit*, trans. Arthur Goldhammer (Boston: Beacon Press, 1984), 13–14.

39 Bachelard, *The New Scientific Spirit*, 13.

40 Bachelard, *The New Scientific Spirit*, 14.

41 Gin S. Malhi and Michael Berk, "Is the Safety of Lithium No Longer in the Balance? [Comment]," *The Lancet* 379, no. 9817 (2012): 691.

42 Baldessarini, *Chemotherapy in Psychiatry*, 96.

43 Baldessarini, *Chemotherapy in Psychiatry*, 96.

44 Bachelard, *The New Scientific Spirit*, 14.

45 R.J.P. Williams, "The Chemistry and Biochemistry of Lithium," in *Lithium: Its Role in Psychiatric Research and Treatment*, eds. Samuel Gershon and Baron Shopsin (New York and London: Plenum Press, 1973), 15–16.

46 Williams, "The Chemistry and Biochemistry of Lithium," 15–16, emphasis added.

47 Williams, "The Chemistry and Biochemistry of Lithium," 15.

48 François Dagognet, "A Regional Epistemology with Possibilities for Expansion," *Science in Context* 9, no. 1 (1996): 6.

9 MOULD

alison kenner
sarah stalcup

Donna Patterson discovered mould in her northeast Philadelphia rental twelve years ago. She was cleaning out a room in her apartment when she found it on a wall behind the heater. The black and green growth covered not only the wall, but also the wood beams that partitioned the sheets of drywall and buttressed them from the floor. Donna worried that the mould had gotten into the heater because of its proximity to the contaminated wall and the amount of mould cover. In response, she stopped using the whole house unit and resigned herself to space heaters until her landlord could take care of the problem. But her landlord did not take care of the problem. Rather, she instructed Donna to clean the mould with bleach – common advice to treat the domestic fungi, but not a tactic recommended by professionals. Donna followed her landlord's instructions but then avoided the room completely. This, by her account, was the beginning of a problem that made her sick, ruined her belongings, and eventually got her evicted.

We met Donna in February 2019 at a workshop on climate change and Philadelphia housing.[1] The workshop was designed to meet the interests of residents who were battling mould and moisture problems in their homes. Moulds are filamentous fungi that inhabit almost every environment on earth. The kingdom "Fungi" emerged approximately a billion years ago when a group of single-celled eukaryotes split into what would become two kingdoms – the kingdom Animalia and the kingdom Fungi.[2] Mould has the distinction of being both macroscopic and microscopic. When people think of mould, they picture green, white, black, and occasionally blue growth on organic surfaces, sometimes fuzzy and sometimes slimy. But what humans see with the naked eye is actually just a glimpse into microscopic communities and life cycles. DNA sequencing has made it possible to characterize different types of mould, as well as mutations in mould communities, but there are thousands of mould species and only a small percentage of moulds have common names. For most home dwellers, there is no way to tell what type of mould you may have, nor its quantity and hazardousness without expert help. Do-it-yourself strategies using home cleaning products are temporary fixes. Even the sage wisdom of eliminating structural nutrient sources – "the taken-for-granted base of our habits and habitats"[3] such as wood and water – is limited in its ability to temper mould's liveliness. Our workshops on climate change present homes as the knotty relations of the Anthropocene, places that exemplify how time, built environments, and capitalism "damage and diminish the ecological envelopes" of domestic dwelling.[4]

In this chapter, we draw on elemental analytics to figure mould as a force to be reckoned with in infrastructure. Mould at home demands our attention, as a life-form that feeds on vital matter, reorders domestic ecosystems, and produces debilitating illness in humans (see figure 9.1). It is, as geographers Sasha Engelman and Derek McCormack describe the elemental, tangible, and excessive of human agency all at once. Mould is known, but usually just in bit and pieces. Documenting mould's mechanisms, habitats, and relationships to other living systems, we ask whether these microbial communities might not be understood as elemental, too, as life forms that interact with other life forms (humans and their environments) to shape objects and agencies.[5] In the case described here, we are particularly interested in how mould inserts itself and exploits well-documented problems with affordable housing, including aged building stock that produces water damage, inadequate legal protection for renters, and absent financial resources needed to root out structural issues, like mould. Not all homes proliferate mould communities but structural inequities seem to nourish these "model ecosystems."[6] In Philadelphia, where the median age of homes is 93 years, the city's poverty rate hovers at 25 per cent, and no one can claim to have a dry basement, a majority of our workshop participants reported attending specifically to learn how to deal with mould in their home.

In this chapter, we reflect on the ways mould is (un)known – through illness, food cultivation, and scientific practice – in order to make sense of the conditions that enable it to proliferate in housing, and to break down homes. Yet the reproductive work of mould communities is not achieved in isolation but rather in tandem with a host of politics.

Something that almost always surprises workshop participants is to hear that we are surrounded by mould, this multiform microorganism, all the time. Residents typically think of mould as an aesthetic, material, or health

Figure 9.1. Mould in bedroom corner, Chicago 2019 (credit: Thomas Anderson / flickr, https://www.flickr.com/photos/senoranderson/4156168701/; CC BY 2.0, https://creativecommons.org/licenses/by/2.0/)

problem isolated to the surface where it appears to the naked eye; but mould is far from the seemingly simple matter that speckle walls and food in our domestic dwellings. For one, mould spores and their communities are a *common* part of indoor and outdoor environments. Our atmosphere is filled with mould spores, though spores cannot be seen the way other atmospheric matter, like pollen or air pollution, might be observed unaided by prosthetic senses like monitors or microscopes.[7] In the US, mould is best known through spore counts, which are measured daily alongside pollen counts at National Allergy Bureau certified stations. Air samplers capture spore concentrations, which allow allergists to calculate the concentration of spores by atmospheric volume. These counts are published daily in air quality reports for each US county, alongside criteria air pollutants and pollen counts, which are more familiar atmospheric features of public health knowledge. Similar to the effects of pollen or pollution for people living with allergies or asthma, high mould spore counts mean that people with mould allergy are at increased risk for symptoms.

Fall is the pinnacle of outdoor mould exposure because mould is a decomposer extraordinaire; the highest mould counts get recorded during this time of year. In order for hyphal cells to grow into mycelium –

the visible material-temporal stage in the fungal life cycle – mould communities require organic material, which is broken down, reabsorbed and recycled when hyphal cells secrete digestive enzymes. Mould communities feast during the fall months as plant material withers, dies, and drops to the ground. As these communities mature, hyphal cells develop spores, which are eventually dispersed, populating the fall air. Nutrients for mould growth are not limited to forest floors or compost bins, of course, but are easily found inside contemporary homes: drywall, carpet, wood beams, but also matter as small as construction dust may also be sufficient at various points in the life cycle. Many moulds also require moisture and temperatures above 4 degrees Celsius (which is why food is refrigerated below this temperature), and some moulds even require light.

Sight and prosthetic sensing are not the only way to identify mould. People who are pathologically sensitized to mould, who have a mould allergy, for example, may sense its presence without seeing it. Some people smell mould, a byproduct of digestive enzymes, perhaps. Mould scents are an active area of research because of what they might tell us about a particular mould community, its secreted enzymes, and mutations.[8] But it is not just about human perception or a matter of assessing whether mould is harmful to us. "Can other microbes sense and respond to the volatiles that mould is making, and is that important for the way that communities of microbes assemble?" asks Dr. Benjamin Wolfe in response to a question about the direction of his research in our 2019 interview. A mycologist who studies microbial communities in food production, Wolfe has found in his research that the answer is yes. Moulds produce volatiles which enable chemical communication within microbiomes: "They're not touching each other, there's no contact, it's just the air that's sending these signals, the volatile signals." This is one of a very few areas of research where scientists are gaining robust understanding of mould species, behaviours, and the evolution of fungi. What they are learning is that chemical signals can change the composition of the microbial community in the local environment. But the chemical communication of mould communities shapes more than neighbouring microbes; it also shapes the conditions of indoor environments, including human occupants.

Shortly after she discovered mould behind her heating unit, Donna became bedridden with a range of symptoms: memory loss, fogginess, diarrhoea, fatigue, and coughing. Her symptoms continued, on and off, for years, making it nearly impossible for Donna to work consistently in her professional occupation as a seamstress. Like most people who suffer from environmental exposures, Donna focused her time and energy on attempts to diagnose her symptoms. None could be traced to an obvious known disease entity. One doctor believed it was a bacterial infection while another thought it might be Lyme disease. Donna received periodic treatments for both. When mainstream healthcare options failed her, she tried healing with complementary and alternative medicine, which included dietary changes, yoga, and meditation. She also received mental health care to help her cope with her debilitating condition and the additional emotional labour that it demanded. Some of her doctors and family members thought Donna's condition was psychiatric in nature and she was eventually diagnosed with bipolar disorder. After years of battling her mysterious condition, she applied for and was granted disability support from the government. Some part of her, though, wondered if her home was making her sick.

Donna showed us more than two dozen pictures when we interviewed her. The walls and wood beams around her heater were peppered with green and black speckled patterns, and also larger blobs of off-coloured watermarks. Her photos extended well beyond the original source of mould – the room with the heater – to include most rooms in her home. Water damage around light fixtures and window sills indicated leaky infrastructure. Containers half-filled with water were randomly placed around her home, positioned carefully to catch rain from a leaking roof. Water feeds mould, of course, allowing mould colonies to not only grow in size and maturity but also to travel through domestic spaces; it is the first thing that one should eliminate when trying to eradicate mould. But this would have been impossible in Donna's apartment given the structural conditions that were producing water damage. Carpets and baseboards highlighted dark, off-coloured blotches, and several pictures showed peeled back wallpaper with similar signs of mould growth. The photo album included evidence of flooding outside the building too, around the foundation of the house and with a blocked drain in front of the garage. One photo showed the photographer's feet immersed in an inch and a half of water at the edge of the garage.

Mould can cause diseases and allergic reactions that range from minor nuisance to life-threatening condition. While mould allergies are well-trod medical territory, mycotoxicosis is much less understood. Moulds feed by secreting enzymes into their environments and then re-absorbing them. It is these enzymes that break down matter like wood, drywall, leaves, and human food products. Some of these enzymes are well-known mycotoxins, "naturally occurring secondary metabolites" produced by toxigenic microfungi.[9] Historically, mycotoxins were observed in relation to food matter and most research today continues to focus on this context.[10] This means that much less is known about the health impacts of mould communities in housing than is known about mould's impact on food production and storage.

Like many environmental illnesses, which are difficult to identify and treat, mycotoxicosis is not among the first things physicians think of when faced with a new patient. The lack of medical science on mycotoxicosis in domestic spaces marks the condition as a contested illness, making it difficult to learn about, receive care for, or pursue legal action if one is made sick by mould.[11] To be sure, mould and their associated mycotoxins are complicated. It bears repeating that mycotoxicosis is not an allergy to mould. Mould allergies are very common, but only impact individuals genetically susceptible to allergic reactions to mould *spores*. As the reproductive component of mould colonies, spores may be carried by water or air, depending on the mould species. Some species that have been linked to allergies are *alternaria, aspergillus, cladosporium* and *penicillium*, but there are thousands upon thousands of mould species, and virtually any mould spore may be capable of producing a reaction in susceptible individuals. Spores may bear mycotoxins, too – this is precisely how people inhale mycotoxins, when they are transported via airborne spores – but the allergy is considered distinct from the toxicity. Unlike allergies, mycotoxicosis operates as a primary pathogen and a natural poison for everyone. The severity and range of symptoms experienced when an individual is exposed to mould toxins depends on such factors as the type of mycotoxin, the amount and duration of exposure, the dietary habits, age, health, and sex of the affected individual. And mycotoxicosis can also increase an individual's vulnerability to other microbial diseases, further compounding health impacts.[12] While it is easy to test for a mould allergy there are no sure tests for mycotoxin.

Unfortunately, Donna told us, none of her doctors suspected an allergy so there was never a referral to an allergist; her symptoms simply did not resemble a classic allergic or asthmatic response. And the idea that Donna might be suffering from a condition many consider to be rare or suspect – mycotoxicosis – never crossed her doctors' minds.

In homes, mould becomes a problem under two conditions: when it proliferates at high density, taking over its environment, or, when the mould is toxic. Bleach continues to be the most widely recommended means to treat mould on indoor surfaces; it is what many of our workshop participants report using in their homes. The problem is, while bleach may kill off and reduce the visible part of the mould community, this chemical treatment does nothing about the mould spores in the air, or those that have landed elsewhere but have not yet germinated. By all accounts, mould spores are the hardiest component of the community. While bleach may keep colonies in check on surfaces, which may be visible because of the mycelium, it can do nothing about decomposition happening underneath the surface, nor spore dispersal. Bleach is toxic, too, an important detail for individuals who are trying to address mould because they are having an allergic or toxic reaction; bleach can have the effect of making respiratory illness or skin reactions worse, for example.

Another approach is to starve mould communities of nutrients, or dramatically change environmental conditions in the home. Eliminating water sources, increasing ventilation, and reducing humidity, for example, are strategies offered by community health workers who help families manage the risks of housing stock vulnerable to mould. These strategies come out of "healthy homes" programming, a form of domestic ecosystem management that seeks to address major public health epidemics like childhood lead poisoning and asthma.[13] In many cases, a combination of reduced moisture and humidity, coupled with surface cleaning can keep domestic mould communities manageable, communities that are by and large nontoxic. But there are also barriers to environmental control, including an inability to ventilate and dehumidify homespaces. For renters, the lack of control over adjacent living spaces may produce vulnerabilities to pests, smoke, allergens, and microbial communities. In Philadelphia's rowhomes, one community health worker told us, problems in one home can easily migrate next door and then down the entire block. Or, as in Donna's case, a leak in the second-floor unit above her, can leach to units below. It is no small task to manage domestic environments, particularly for those who live with illnesses that are triggered by microbial ecosystems.

Not everyone wants to get rid of mould, of course. And even if you could get rid of mould – which you cannot – it would be wholly detrimental to living systems. Permaculturists and ecologists, for example, recognize mould as critical for healthy environments. Cheesemakers, too. Purchasing the right mould culture can be essential for making Brie and Camembert, for example, and so cheesemakers must carefully regulate environments to cultivate and control for specific microbial communities.[14] This is a tall order for a range of reasons, not the least of which is our limited ability to identify mould species, know the maturity of mould communities, and track their evolution pathways. In a 2019 interview with a scientist who works with cheesemakers, we discussed the challenges of caring for production environments. The scientist noted that cheesemakers' work, too, relies on an ability to perceive and know how to manage visible moulds:

In many ways [cheesemakers] are gardeners that are wearing blindfolds. They have to cultivate particular communities of organisms, but they can't see them. They see some sign, so they see they're Camembert is getting white fuzz on it, but in that white fuzz there's a whole jungle of other things that they can't see or don't know exactly what's there. It's a problem for them in many ways because they have really blunt tools and often they're blinded. They often don't know who's coming, who's going, who's changing their environment. I think actually the easiest part to manage and the easiest signals of change are the moulds because they can see those, they can actually physically see a black mould coming in on their cheddar, or the lack of mould. Those are the sort of microbial barometers for them, which allow them to understand that there are shifts happening in their system.

Our sense of mould – whether by smell, sight, or mould count – gives us some information about the environment in relation to human needs (food, shelter surface, breathing space), but our tools for apprehension and management are blunt, even among experts. Mould in the colloquial sense is tangible, but it is only a part of much larger microbial communities that far exceed human intervention.[15]

If elements have formerly been thought of in terms of opposing poles – as unitary substance or as environmental milieu – one innovation in elemental thinking may be to think in terms of communities. What is seen on the surface of a ceiling, a shower tile, or a piece of fruit belies a structure of reproduction, support, and communication that exceeds quotidian perception, and the regimes that regulate domestic environments.

Today, Donna is convinced that the condition of her home environment – water damage that led to mould overgrowth – is what made her sick for so many years. Yet as a renter, there was only so much Donna could do to address the dilapidated state of her dwelling. For years, she asked her landlord to fix leaks and drainage problems, and to update appliances – to no avail. From Donna's description, there was a long, symbiotic relationship with her landlord that made it difficult to navigate maintenance issues. When she moved into the unit twenty years ago it was ideal because there was enough space to conduct her work as a seamstress. During her long illness, Donna got behind on her rent, but her landlord gave her some relief until she could get back on her feet, which she did eventually. But the price of back-rent forgiveness was that Donna needed to take on maintenance responsibilities. She was able to make some repairs herself but, over time, many of the problems surpassed her resources and Donna had long absorbed the costs of deferred maintenance. By fall 2018, two decades of structural neglect had accumulated and most of the needed repairs were beyond her means. She had to do something to get her landlord to respond. Moving out was not an option. This was her home, she told us.

Based on the advice of friends, Donna pursued the theory that her apartment had a mould problem. She hired a contractor to do a mould test, and sure enough, the contractor found extensive mould contamination, including black mould and three other dangerous mould species. Donna reported her landlord to License and Inspections, a Philadelphia municipal agency that governs the safety of properties, and they indeed ordered her landlord to repair the structural issues that Donna had documented – leaks from the old roof and appliances, drainage problems, and extensive mould growth behind wallpaper and underneath carpets. But in response, Donna's landlord evicted her with only thirty days' notice. The stated reason was "illegal operation of a business." "That's when I started studying the law instead of mould," Donna told us. "I was trying to make her do the right thing and fix it. And I discovered, you can't make an illegal landlord do the right thing, under your thumb."

Donna's story is one we hear often in Philadelphia. Residents hear such stories too and, thus, landlord retaliation is a commonly reported reason renters continue to live in housing conditions that may be making them ill. Existing laws do not prevent eviction in such cases, and gaps in the legal system cause people like Donna to become homeless. If in the early years of the twenty-first century mould had been deemed the "new asbestos," to quote the US International Trade Commission, the insurance industry quickly pivoted to make sure that mould damage and remediation exceeded what was covered in most policies.[16] Mould lawsuits that have been won were the result of financial and legal resources wholly out of reach for Donna Patterson and those living in similar conditions.

Regulatory infrastructure depends on knowledge, and knowledge of mould is complicated and patchy. In a culture that privileges what is visible, our everyday relationships to microscopic matter remain uncertain and contested in medical and legal arenas. Physical, visible evidence must accumulate before knowledge claims are made, and thus action taken. Such is the case with air pollution or chemical exposures.[17] But unlike outdoor pollution and some industrial toxics, which are governed by environmental laws that protect public health (however poorly), there is no regulatory apparatus for mould, as Donna found out. When a resident faces mould contamination in the state of Pennsylvania, there is no agency or law to appeal to. In the absence of medical and scientific knowledge claims that could inform laws for residential environments, home dwellers rely on intimate, sensorial practice to care for domestic spaces.[18]

Mould has become politicized in the last twenty years because of how it is emerging within the intimate spaces of our home environments. As a force of new consequence amid changing climates, aged housing stock, and inadequate legal apparatus, microbial communities make domestic spaces lively, if on uncertain terms. While the mould you come into contact with in everyday practices remains innocuous, the potential for new forms of life, new communities, is ever present. Imperceptibly pervasive, mould's potential as microbial community materializes in an emergent sensorial politics of dwelling, reliant upon attunement to smells, humidity, and embodied signals of contact with allergens. Whether in food production, pharmaceutical development, or even in managing home environments, human effort to control mould will almost always be stifled by its excess, endurance, and adaptability.

NOTES

1 To learn more about the Climate Ready Philly workshops, visit https://disaster-sts-network.org/content/climate-ready-philly-0/essay.

2 Jason E. Stajich et al., "The Funghi," *Current Biology* 19, no. 18 (2009).

3 John Durham Peters, *The Marvelous Clouds: Toward a Philosophy of Elemental Media* (Chicago, IL: University of Chicago Press, 2015), 1.

4 Neale, Addison, and Phan, "Introduction," this volume.

5 Heather Paxson and Stefan Helmreich, "The Perils and Promises of Microbial Abundance: Novel Natures and Model Ecosystems, from Artisanal Cheese to Alien Seas," *Social Studies of Science* 44, no. 2 (2014).

6 Paxson and Helmreich, "The Perils and Promises."

7 Joseph Masco, "Nuclear Technoaesthetics: Sensory Politics from Trinity to the Virtual Bomb in Los Alamos," *American Ethnologist* 31, no. 3 (2004).

8 Benjamin E. Wolfe et al., "Cheese Rind Communities Provide Tractable Systems for In Situ and In Vitro Studies of Microbial Diversity," *Cell* 158, no. 2 (2014); Benjamin E. Wolfe and Rachel J. Dutton, "Fermented Foods as Experimentally Tractable Microbial Ecosystems," *Cell* 161, no. 1 (2015).

9 Estelle Levetin et al., "Taxonomy of Allergenic Fungi," *The Journal of Allergy and Clinical Immunology: In Practice* 4, no. 3 (2016).

10 J.W. Bennett and M. Klich, "Mycotoxins," *Clinical Microbiology Reviews* 16, no. 3 (2003).

11 Phil Brown, *Toxic Exposures: Contested Illnesses and the Environmental Health Movement* (New York, NY: Columbia University Press, 2007).

12 Bennett and Klich, "Mycotoxins."

13 Alison Kenner, *Breathtaking: Asthma Care in the Time of Climate Change* (Minneapolis, MN: University of Minnesota Press, 2018).

14 Heather Paxson, *The Life of Cheese: Crafting Food and Value in America* (Berkeley, CA: University of California Press, 2012).

15 Sasha Engelmann and Derek McCormack, "Elemental Aesthetics: On Artistic Experiments with Solar Energy," *Annals of the American Association of Geographers* 108, no. 1 (2018).

16 Industry and Trade Review, 2005.

17 Max Liboiron, Manuel Tironi, and Nerea Calvillo, "Toxic Politics: Acting in a Permanently Polluted World," *Social Studies of Science* 48, no. 3 (2018).

18 Liboiron, Tironi, and Calvillo, "Toxic Politics."

10 MYLAR

derek p. mccormack

One of the challenges of thinking about anthropogenic elements is the fact that there already are many versions of the elements in circulation across different domains of experience and expertise.[1] The elements can be understood as ontological propositions about the composition of bodies and worlds. This is the version found in various classificatory schemes about the fundamental categories or roots of matter, categories which typically include, without being limited to, air, water, earth, fire, wood. The elements can also be understood, more narrowly, as molecules and compounds, especially if framed via the epistemic reductivism of the periodic table.[2] They can be understood as forms of environmental surrounds or milieus.[3] And they can be understood as media.[4] None of these versions exists in isolation – each circulates in relation to the others in complex ecologies of association, contamination, and implication. Each version ghosts the others, even when different versions are emphasized in particular circumstances.

To focus on anthropogenic elements is to add another version to this mingled elemental given.[5] Concentrating critical attention on one kind of anthropogenic object or material is not however the same as isolating one version of the elements. Instead, the constraint provided by this attention facilitates possibilities for thinking across different versions of the elements and their participation in the composition (and decomposition) of forms of life. It also affords distinctive genres for tracing versions of the elements as they shape the (de)generation of worlds. These genres can include various practices of thing-following or object-biographies but they can also involve tracing how anthropogenic elements shape-shift as materials in process in relation to different conditions of life and death.

Focusing on film provides one constraint and genre, cutting as it does across different versions of the elements. Most obviously perhaps, film foregrounds the importance of thinking about elemental *media* in relation to the other versions of the elements just outlined. Film is a medium through which to experiment, for instance, with technologies of perception and affective forms in ways that render visible and palpable the elemental materiality of milieus like water and air. Film's capacities in this respect go well beyond the representation of elemental phenomena or experiences. It extends to the invention of images that generate forms of "gaseous perception" not centred on any subject, but defined instead by the "free movement of each molecule."[6] In this way film folds together media, milieu, and molecule, blurring, in the process, neat distinctions between these versions of the elements. And it does so in ways that both complicate and extend thresholds of sensing and perception.

Film is by no means the only form of elemental media through which to explore how the conditions of bodies, forms of life, and worlds are transforming in the wake of the multiple hyper-objects of anthropogenic activity. Film is especially interesting, however, in relation to the kinds of elemental exposures generated by these activities. As an assemblage of image-making technologies, practices, and experiences, film has of course always been about managing and modifying elemental exposure. In its analogue incarnation, film is a material on which images are recorded, and film-making is a photochemical process involving the exposure to light of an emulsion on a celluloid or polyester base. This understanding of film, as a material of exposure, points to a much older and wider meaning of the term. The English word "film" is derived from the Germanic term *filmene,* meaning a kind of skin or membrane. At times referring to specific parts of the human body, film can also describe a thin layer over the eye of a dying person obscuring their vision or a similarly obscuring layer generated by an intense emotional experience. Its association with the clouding of perception extends into a more general understanding of film as a thin layer of "haze, mist, or other obscuring feature."[7] In these terms the earth's atmosphere can be grasped as a film emanating from the planet, modifying the exposure of other bodies to the elements and being itself modified by the anthropogenic production of elements from carbon to sulphur. The orbit of film – as technology and material of envelopment and exposure – therefore stretches well beyond practices of image-making to encompass multiple versions of the elements and their earthly implications.

Understood in these expansive terms film is a category that includes anthropogenic materials used for wrapping, containing, enveloping, and shielding. These kinds of materials are neither new nor distinctively anthropogenic. However, central to the many anthropocenes in relation to which life and death take place today are distinctive versions of these materials.[8] Products of twentieth century industrial chemistry, polymeric films are layered into ordinary worlds in the guise of polyethylene, polypropylene, polyester, nylon, polyvinyl chloride, cellulose acetate, and cellophane. They are used in packaging, labelling, bagging, clothing, and coverings, while also providing base materials for audio and video recordings. These synthetic films can of course be defined rather reductively in molecular terms: their production involves, at its most basic, the manipulation and arrangement of chemical elements. But they also implicate other versions of the elements, and particularly so because as materials of envelopment, plastic films modify the exposures of bodies – human and non-human – to the "force" of elemental agencies and milieus.[9] To think about films as anthropogenic elements therefore involves exploring how molecules are fashioned into materials that alter relations between bodies and their surrounds via distinctive forms of being enveloped and being exposed.

Developed and manufactured by DuPont, Mylar is one of the trade names of biaxially oriented polyethylene terephthalate (or BoPET). This clear film is formed by extruding a sheet of molten polyester onto a chilled surface and then drawing it in one direction and then another. Trademarked by DuPont in 1952, Mylar has a high tensile strength, is dimensionally stable (resistant to stretching), chemically neutral, non-yellowing, non-tearing, electrically insulating, and highly impermeable to moisture, gas, and odour. With an operational temperature range of about -100°F to 300°F, Mylar film is used in packaging, printing, electronics, protective wrapping, clothing, and food preservation.

Mylar, like other polymer films, can be metallized by the application of a thin coat of evaporated aluminium in a process that produces a light, flexible, highly reflective material. This version of Mylar is often associated with helium-filled party balloons. In fact, these balloons are not usually made from Mylar. However, just as the brand name "Hoover" has become a generic term for vacuum cleaners so too has Mylar become a generic term for any metallized plastic film, even when the base polymer is a different material. Mylar simultaneously denotes a specific anthropogenic element and a wider set of anthropogenic materials forming a lightweight substrate of what Mimi Sheller calls "light modernity," organized around the interwoven technical promises of lightness and strength.[10] Combined as it is with a vaporized version of the element about which Sheller writes – aluminium – metallized Mylar has a distinctive reflective allure that became bonded into the promissory fabric of the chemically enhanced good life of post-war liberal democracies.[11] DuPont was at the forefront of the production of promotional films through which this bonding took place: in these films materials like Mylar were presented as wondrous solutions for problems that consumers had yet to realize existed. One commercial from 1954 begins with a baseball player pitching a ball at the viewer. It bounces off. The trick is then revealed: a thin film of Mylar had been placed in front of the camera. The camera then pans to a domestic interior, where future uses of the material are highlighted, including clothing, audio tape, insulation, and wall coverings. The ad ends with a woman swinging from Mylar strands suspended from a tree.

The promise of Mylar's properties and the worlds in which it could become a transformative material were more elaborately staged in a 20-minute film, *What's It to You?* (see figure 10.1).[12] Echoing the shorter ad, this one also opens with a sharp-suited narrator hitting a thin sheet of Mylar with a baseball bat. *What's It to You?* illustrates in detail the molecular composition, practical properties, and aesthetic allure of Mylar. Viewers are invited to look "at an artist's conception of a Mylar molecule" and to note how, in the process of polymerization, "the arrangement of molecules is changed from a helter skelter pattern ... to a more orderly arrangement in which the long chains are firmly fitted together, giving the film unusual properties." "Yes," as the narrator claims, "Mylar's properties are right in its molecules." These properties are then illustrated through a range of activities. "Tensile strength" is demonstrated by a male acrobat bouncing vigorously on a Mylar trampoline; "resistance to tearing" by a female acrobat swinging from a sheet of Mylar twisted into a rope; and "impact strength" by a bowling ball colliding at "full speed" with a Mylar sheet. A bag fashioned from Mylar containing a "strong solution of nitric and sulphuric acid" is shown. Female assistants wear the material, wipe it down, and in one case hold up a live skunk enveloped in Mylar to demonstrate the material's impermeability to odour. And a loop of Mylar is passed through dry-ice and steam while a recording of the narrator's voice continues to play, undistorted. The film ends by amplifying DuPont's corporate refrain of "better living through chemistry":

> The wide range of properties of Mylar allows us to explore a whole world of ideas we never dreamed existed ... More and more the properties of this new polyester film are bringing new ideas to industry and the home, to manufacturing and distribution, and with them are coming new and better things to us all. What's it to you? As we said, a good question. With Mylar, you pick your own answer.[13]

Figure 10.1. Still from *What's It to You?* (credit: DuPont, 1955)

One of the questions for which Mylar provided an answer was this: How to modify the exposure of bodies to the conditions of their elemental milieus? As packaging, Mylar film can be used to slow rates of chemical decomposition and to facilitate the production of modified/controlled atmospheres (especially important for the shelf-life of processed food). Critically, in metallized form it can also be used to maintain temperature differentials between bodies and their surroundings because such films "virtually eliminate radiant heat transfer."[14] Managing and modifying heat transfer became a critical part of the infrastructures of elemental extraction and use in the early twentieth century. This was especially important in cryogenic research into the production, storage, and transportation of liquefied gases including oxygen, helium, and hydrogen. Such gases were essential to the cooling systems that facilitated the development of both the US atomic weapons and space programs. Metallized films were particularly important in the US space programs, offering critical thermal protection for probes, spacecraft, and the bodies of astronauts. These films provided a vital thermal layer in space suits, especially those worn during extravehicular activities (EVAs), the first of which was undertaken in 1965. Such suits were composed of different materials, many of which, including Mylar, Teflor, Dacron, Nylon, Neoprene, Nomex, Kapton, and Lycra, were manufactured by DuPont.

Beyond their functionality, these metallized films were also often highly visible in ways that accentuated their aesthetic allure, and especially on spacecraft. One of these metallized films, Kapton, was used to fabricate the yellow-gold reflective blankets on the lunar modules of the Apollo missions which, alongside Mylar, were critical to managing the thermal conditions of these craft and their technical operation. A polyimide rather than a polyester aluminized film, Kapton can withstand temperatures from close to absolute zero to 400C. As DuPont reminded consumers in its advertising, Kapton was the first material on the Apollo spacecraft to make contact with the Lunar surface.[15] The layers of crinkled Kapton with which the base and legs of the landers were enveloped gave a shimmering lustre to these craft, while also suggesting a homemade hand-crafted fragility. The lightness of these materials had other consequences when the astronauts departed the surface of the moon, however. Footage of the departure of Apollo landers taken from cameras on the lunar surface show how, just after the moment of ignition and lift-off, fragments of both films radiated at high speed in all directions like glittering confetti. The process was observed by both Neil Armstrong and Buzz Aldrin, and described in detail by the latter who suggested they "seemed almost to be going out with a slow-motion type view [in] enormous distances from the initial PYRO firing."[16]

Aldrin's observation is a further reminder that relations with metallized films are always more than technical: the reflective and scintillating qualities of these films also generate distinctive aesthetic experiences that, while mirroring Earth-bound "glitterworlds,"[17] seem to take on a particularly spectral, otherworldly quality in space. During the Apollo 10 mission, the crew of the command module made a photographic pass over the proposed landing site for Apollo 11 before moving into the dark side of the moon. On returning to the sunlit side, astronaut John Young reported seeing "maybe a foot-and-a-half or more of Mylar with that insulation coating on the back of it." Appearing "out in front of [the] window," Young noted that the Mylar "just sort of sat there for a while, and then quietly floated off."[18] Young's concern with the presence of the object was primarily about possible risks to the command module but there is still a hint in his description of the uncanniness of an encounter with an anthropogenic element becoming otherworldly. And there is a hint of the kind of allure implicated in the process of that element withdrawing from the perceptual orbit of its origins. This sense of allure can be and has been amplified in encounters with other fragments of metallized films. In 1998, during an EVA as part of the construction of the International Space Station, a piece of Mylar thermal insulation was lost by one of the astronauts. Photos of it were captured by astronauts on the Space Shuttle *Endeavour*.[19] These images started to circulate within conspiracy theories about UFOs and alien visitors, resonating particularly intensely with stories of the "Black Knight Alien Satellite" that centre on the possible existence of an ancient extra-terrestrial craft in orbit around the earth.

The allure of metallized films and their association with new experiences of space and encounters with objects was also something with which many artists experimented in the 1960s, inspired in part by the space program. Objects crafted with or covered by Mylar and other films could, when filled with helium, float and drift with a waywardness that seemed to echo both the experience of bodies operating under different gravitational conditions and the production of new kinds of anthropogenic foreign bodies. Notably, Andy Warhol's *Silver Clouds* (1966) used Scotchpak (a metallized laminate of polyethylene) envelopes filled with helium and air to create an installation of floating pillows that moved in response to the movement of air in a gallery space. Warhol was enthusiastic about the astronautical associations of the material, claiming that "silver was the future, it was spacey – the astronauts wore silver suits – Shepard, Grissom, and Glenn had already been up in them, and their equipment was silver, too."[20] And his *Silver Clouds* anticipated the kinds of experiences with floating pieces of Mylar in space later recounted by astronauts like Young.

The light reflectivity of Mylar proved alluring for other artists in the 1960s. Les Levine (nicknamed "plastic man") frequently used synthetic films as part of disposable art and in the creation of what he called "environmental places."[21] *Windows* (1969) consisted of thousands of die-cut aluminized Mylar stickers that could be sold and disposed. *Slipcover*, installed at the Walker Gallery in 1967, consisted of a room covered with aluminized Mylar in which two large envelopes of the same material gently inflated and deflated.[22] *Slipcover* was intended by Levine to emphasize the experience of envelopment and immersion, in a space that moved and stretched the perceptual and kinetic reference points of phenomenological experience. Part of the appeal of Mylar in such experiments was how its reflectivity afforded opportunities for experimenting with images, exemplified in the fabrication from metallized film (Melinex) of the Mirror Dome of the Pepsi Pavilion in 1970.[23] Equally, during a period from 1968 until 1971, artist and photographer Ira Cohen experimented with Mylar in the production of photographic images that seemed to accentuate the altered perception of psychedelic experience.[24]

In such experiments metalized films, shaped into floating or breathing objects, work to disorient experiences of immersion in elemental milieus. Nevertheless, they arguably remained organized around the sensory thresholds, however distorted, of a distinctly human subject of perception. Equally, foregrounding the spectral allure of metallized film in such works needs to be tempered by the fact that the scattering and dispersal of this film is part of the emanation of a cloud of anthropogenic waste materials on and beyond Earth whose scale and form is too vast to grasp from the perspective of the individual subject. Other artists have used Mylar in ways that foreground the accumulation of non-human entities and objects in ways that gesture to this. Exemplary here is Tara Donovan's *Untitled, Mylar* (2010) which consists of a large number of spheres of different sizes, each formed of sheets of tightly folded Mylar.[25] The spheres are arranged in a sculptural form that looks like an enormous molecule of black carbon dust with reflective facets. As in much of Donovan's work, *Untitled, Mylar* uses ordinary materials to generate forms that transcend the scales of their component objects.[26]

What the stark, fixed, apparent solidity of Donovan's piece does not really show is the capacity of a metallized film like Mylar to move, drift, and disperse. Nor does it really generate a sense of the fact that metallized films are often encountered, by humans and others, as isolated fragments rather than in aggregate.[27] Other work captures this sense of fragmentary scattering and dispersal. *Untitled (Mylar)* (2011) is a short film by Brooklyn-based artist William Lamson. It features a sheet of aluminized Mylar being blown by the wind across a dried lake bed (see figure 10.2). The sheet is tracked by a camera which itself is moving at the same speed as the Mylar. Never gaining altitude, the Mylar stays low, brushing the ground. *Untitled (Mylar)* exemplifies Lamson's wider interest in experiments of different duration with the generative possibilities of elemental processes including emergence, submergence, crystallization, and precipitation. In this short video Mylar itself becomes a film in motion, continuously shape-shifting, folding, and refolding through a topological process in which different versions of the elements participate.

Lamson's film is deliberately staged – a camera on the back of a truck tracks the blanket as it is blown across a desert floor after having been released by the artist. And yet it is not at all far-fetched to imagine that such windblown materials could be encountered in this way, especially given their disposability and discardability. After all, these blankets are in common use in many situations. These uses include the regulation and modification of the exposure of human bodies in relation to the thermal conditions of their surroundings. Originating in the space industry, these blankets have now become central to what Nicole

Figure 10.2.   Still from *Untitled (Mylar)* (credit: William Lamson, 2011)

Starosielski calls "thermocultures" associated with practices as diverse as marathon running, hiking, and mountaineering, and survivalist practices of preparedness.[28]

The blanketing of bodies in metallized films is also one of the signature gestures and images associated with the aftermaths of various kinds of disaster. It is one of the generic forms in which emergency response is materialized in ways that seem to give shape to a highly choreographed imperative to care. In 2005, for instance, a "Heatsheet Thermal-Blanket Initiative" saw 120,000 of these blankets distributed by MercyCorps to victims of an earthquake centred near Balakot, Pakistan.[29] In such cases, the Mylar blanket is humanitarianism rendered immediate in the shape of a thin veneer of comfort. But the very same gesture of blanketing bodies in Mylar can also be a sign of how other bodies are not considered worthy of care or are themselves always at the threshold of disposability or discardability. Indeed, in certain circumstances the distribution of these blankets and their envelopment of particular bodies seems to reflect the exposure of these bodies to the political and affective conditions of hostile environments. In 2016, images of immigrants detained in holding cells in Arizona were released as part of a lawsuit against the US Customs and Border Protection Agency. The images showed people crowded in concrete cells, often without bedding, and surrounded in many cases by trash. Notably, in many of the images detainees were wrapped in Mylar blankets. In one image, taken in a holding facility in

Tucson, detainees were pictured cocooned in Mylar blankets while lying on the floor of a cell.[30]

The same reflective properties that provide a thin layer of thermal envelopment in conditions of violence can also be used to contest such conditions. Mylar blankets have been used as devices for protesting against the detention of immigrants and the conditions in which they are detained. Such protests capitalize on the alluring reflectivity of the blankets. In June 2018 about 70 children demonstrated against the detention of immigrants and the conditions of their detention by wrapping themselves in Mylar blankets and staging a protest on the floor of the Russell Senate Office Building on Capitol Hill in Washington, DC.[31] Another protest, also in June 2018, ended with many demonstrators, again wrapped in Mylar, occupying part of the Hart Senate Office Building on Capitol Hill.[32] Similar gestures have been used to highlight the conditions of refugees trying to cross the Mediterranean from North Africa to Southern Europe. In 2016 the artist Ai Weiwei distributed Mylar blankets to guests, including celebrities like Charlize Theron, attending a charity event in Berlin which had been organized to raise awareness of the plight of these refugees.[33] Weiwei's choice of these blankets was informed by his awareness of their association with images of people adrift and dying in the Mediterranean. His creation of the spectacle of a room of celebrities wearing Mylar received some criticism, not least because distributing the blankets in such a context seemed to further aestheticize the traumatic experiences to which the event gestured. Nevertheless, in part because of the way they draw they eye while also suggesting the spectrality of discarded bodies, Mylar blankets have come to figure prominently in artistic engagements with "migration or mobility, alongside other ghostly figures in transit."[34] For instance, in order to foreground the association between these blankets, experiences of displacement, and the condition of being discarded, the artist James Bridle used a piece of Mylar to construct a temporary installation at Ellinikon in Greece in a work named *A Flag For No Nations* (2016).

These artistic deployments are minor acts of creative repurposing that draw attention to the politics of disposability. Other minor acts of repurposing metallized film are also possible. For instance, forms of "stealth wear" crafted from metallized film are part of tactics to operate below the thresholds of various assemblages of visual and thermal surveillance.[35] Metallized film modifies the thermal signature of bodies in ways that mean they do not show up so clearly as images on the screens of the sensing devices of different formations of power.[36] These capacities of metallized films figure prominently in the worlds of US-based preppers anticipating pre- and post-apocalyptic state surveillance.[37] In a very different context, bodies shielded by Mylar also figured during the 2019 protests in Hong Kong. Reports suggested that demonstrators on the street sometimes wore Mylar blankets to evade detection by thermal surveillance cameras deployed by security services. That is not to overstate the significance of these tactics. In the case of Hong Kong, for instance, stories of these tactics mingled with images circulating of emergency services distributing blankets to protestors who had tried to escape from the Polytechnic University campus through a sewage pipe.[38]

The broader point here is that such situations disclose how bodies are blanketed by metallized films in ways that reveal their relative vulnerability and exposure to the violence of worlds.[39] These situations reveal how metallized film is at once the routine manifestation of the minimal care prescribed by the protocols of emergency response and, at the very same time, reflective of the participation of this response in regimes that consider the lives enveloped by these films to be disposable. Mylar blankets are gestures towards the bare, technical modification of elemental relations within milieus that might keep bodies living without having to address the systemic and structural violence that renders those milieus increasingly inhospitable.

Living and dying with anthropogenic elements is in no small part a matter of the relations between conditions of envelopment and exposure. Film is an anthropogenic element for modifying these relations in ways that implicate many versions of the elements. The emergence and experiences of the COVID-19 pandemic during the writing of this chapter only served to intensify this. In the midst of this unfolding event, plastic films of various kinds were repurposed as part of many different techniques and practices of envelopment and exposure. Many retail employees worked behind screens of these materials. Meals were eaten "together" on different sides of these films. Air and atmospheres were partitioned and compartmentalized. And people embraced and hugged while separated by thin layers of film.[40]

This last image, of bodies impressing themselves on each another through films while minimizing exposure, is perhaps especially evocative. It also echoes a much earlier formulation of elemental media. In Book IV of *De Rerum Natura*, Lucretius conjures a material phenomenology of sensation in which images, some too fine to be perceived, emanate from bodies. He writes:

> These images are like a skin, or a film,
> Peeled from the body's surface, and they fly
> This way and that across the air; they cause

A terror in our minds, whether we wake
Or in our sleep see fearful presences.
The replicas of those who have left the light
Haunt us and startle us horribly in dreams.

...

Let me repeat: these images of things,
These almost airy semblances, are drawn
From surfaces; you might call them film, or bark,
Something like skin, that keeps the look, the shape
Of what it held before its wandering.[41]

Lucretius is writing about human bodies as sites and sources of elemental perception. These films are images in the Bergsonist sense: they have an elemental materiality that lies somewhere between a representation and a thing, their allure constituted in part by the fact that they never resolve themselves into either.[42] Even if archaic, Lucretius's ideas provide some orientation for thinking of film as one of the products of the multiple anthropocenes in relation to which the present is tensed. The chemical engineering of synthetic films, including Mylar, is one of the processes by which anthropogenic forms of life extrude, shed, and discard materials as part of the engineering of elemental worlds.[43] These films modify the exposures of bodies – human and non-human – to the conditions (thermal and otherwise) of different elemental milieus. They can work to shelter and shield bodies, taking the momentary form of those bodies in the process. But they also appear as shrouds of a kind: layers obscuring bodies in the wake of their being abandoned by different political circumstances. These are films that promise comfort while also revealing exposures to toxicity and violence. Films wandering, flying this way and that, scattering across surfaces as diverse as the Mediterranean and the Moon – spectral presences persisting as the shape of lives being enveloped and discarded.

NOTES

1 Sasha Engelmann and Derek McCormack, "Elemental Worlds: Specificities, Exposures, Alchemies," *Progress in Human Geography* (2021).

2 Michelle Murphy, "Alterlife and Decolonial Chemical Relations," *Cultural Anthropology* 32, no. 4 (2017).

3 Astrida Neimanis, "The Sea and the Breathing," *E-flux*, accessed 24 May 2020, https://www.e-flux.com/architecture/oceans/331869/the-sea-and-the-breathing/.

4 John Durham Peters, *The Marvelous Clouds: Toward a Philosophy of Elemental Media* (Chicago, IL: University of Chicago Press, 2015); Nicole Starosielski, "The Elements of Media Studies," *Media+ Environment* 1, no. 1 (2019).

5 Michel Serres, *The Five Senses: A Philosophy of Mingled Bodies*, trans. Margaret Sankey and Peter Cowley (London: Bloomsbury, 2008).

6 Gilles Deleuze, *Cinema 1: The Movement Image* (London: Bloomsbury, 2005): 86.

7 *Oxford English Dictionary*, https://www.oed.com.

8 Alison Kenner, Aftab Mirzaei, and Christy Spackman, "Breathing in the Anthropocene: Thinking through Scale with Containment Technologies," *Cultural Studies Review* 25, no. 2 (2019).

9 Peter Adey, "Air's Affinities: Geopolitics, Chemical Affect and the Force of the Elemental," *Dialogues in Human Geography* 5, no. 1 (2015).

10 Mimi Sheller, *Aluminum Dreams: The Making of Light Modernity* (Cambridge, MA: MIT Press, 2014).

11 Rebecca Altman, "American Petro-topia," *Aeon* 11, March (2015); Jeffrey Meikle, *American Plastic: A Cultural History* (New Brunswick: Rutgers University Press, 1997); Jennifer Gabrys, Gay Hawkins, and Mike Michael, *Accumulation: The Material Politics of Plastic* (Oxford: Routledge, 2013); Heather Davis, "Life & Death in the Anthropocene: A Short History of Plastic," in *Art in the Anthropocene: Encounters Among Aesthetics, Politics, Environments and Epistemologies*, eds. Heather Davis and Etienne Turpin (London: Open Humanities Press, 2015): 347–58.

12 DuPont DeNemours and Co., Inc., *What's It to You?* (Chicago, Jam Handy Films, 1955).

13 DuPont DeNemours and Co., Inc., *What's It to You?*

14 Midwest Research Institute, *Reflective Superinsulation Materials. Report No. CR-2507*, Prepared for NASA (Washington DC: NASA, 1975), 2. Available at https://ntrs.nasa.gov/archive/nasa/casi.ntrs.nasa.gov/19750006837.pdf.

15 DuPont, "Apollo 11: How Partnership Prepared Us for Space," 18 July 2019, accessed 15 December 2010, https://www.dupont.com/news/innovation -and-partnership-prepared-us-for-space.html.

16 NASA, "Apollo 11 Technical Crew Debriefing," 31 July 1969 (Houston, TX): 89.

17 Rebecca Coleman, *Glitterworlds: The Future Politics of a Ubiquitous Thing* (Cambridge, MA: MIT Press, 2020).

18 Apollo 10 Digital Picture Library, *Apollo Lunar Surface Journal*, accessed 21 December 2019, https://history.nasa.gov/alsj/a410/images10.html.

19 Gateway to Astronaut Photography of Earth, accessed 21 December 2019, https://eol.jsc.nasa.gov/SearchPhotos/photo.pl?mission=STS088&roll =724&frame=66. See Daniel Oberhaus, "Alien Hunters Spent the Last Century Looking for the Black Knight Satellite," *Vice*, 2015,

https://www.vice.com/en_us/article/4xagnb/alien-hunters-spent-the-last
-century-looking-for-the-black-knight-satellite; Cara Giaimo, "The Weird
History of the Space Blanket," *Atlas Obscura*, 2016, https://www.atlasobscura
.com/articles/the-weird-history-of-the-space-blanket.

20 Superstars & Silver: Factory Photographs 1964–1970, accessed 21 December 2019,
https://www.bbc.co.uk/programmes/articles/10dt7NbCg1bbVtWL08Wsy6v
/superstars-silver-factory-photographs-1964–1970.

21 Meikle, *American Plastic*, 238.

22 Pavel Pyś, "Second Thoughts: Les Levine at Walker Art Center (1967),"
Sightlines (Minneapolis, MN: Walker Art Center, 2016).

23 Fred Turner, "The Corporation and the Counterculture: Revisiting the
Pepsi Pavilion and the Politics of Cold War Multimedia," *The Velvet Light
Trap* 73 (2014): 66–78.

24 Ira Cohen, *Into the Mylar Chamber* (Somerset, UK: Fulgar Press, 2019).

25 Timothy Morton, *Hyperobjects: Philosophy and Ecology after the End of the World*
(Minneapolis, MN: University of Minnesota Press, 2013).

26 Indianapolis Museum of Art, "Tara Donovan's *Untitled (Mylar)*," in *Smarthistory*,
May 9, 2017, https://smarthistory.org/tara-donovans-untitled-mylar/. See
also Amanda Boetzkes, *Plastic Capitalism: Contemporary Art and the Drive to Waste*
(Cambridge, MA: MIT Press, 2019).

27 Julia Sizek, "Pollution Does Not 'Go Away': Mylar Balloons and Air Quality in
California's Pristine Wastelands," *Engagement*, 2019, https://aesengagement
.wordpress.com; John Lindsey, "Why Death Valley Is Littered With Mylar
Balloons," *The San Luis Obispo Tribune*, 14 June 2014.

28 Cory Lundin, *When All Hell Breaks Loose: Stuff You Need to Survive When Disaster Strikes* (Salt Lake City: Gibbs Smith, 2007); Nicole Starosielski, "Thermocultures of Geological Media," *Cultural Politics* 12, no. 3 (2016).

29 See MercyCorp, "Pakistan Earthquake Response Heatsheet Thermal-Blanket
Initiative Final Report, 18 May 2006, https://www.mercycorps.org/files
/file1149543181.pdf.

30 Camila Domonoske, "Surveillance Stills from Border Patrol Facilities Show
Crowded, Trash-Filled Cells," *The Two-Way, NPR*, 19 August 2016.

31 Ali Rogin, "Children Don Mylar Blankets, Protest Family Separations on Capitol
Hill," *ABC News*, 21 June 2018, https://abcnews.go.com/Politics/children
-don-mylar-blankets-protest-family-separations-capitol/story?id=56057267.

32 Jenni Fink, "Anti-Ice Protest: Women Arrested In Senate Building While
Protesting Donald Trump's Immigration Policy," *Newsweek*, 6 June 2018.

33 Henry Barnes, "Celebrities Don Emergency Blankets at Berlin Fundraiser
for Refugees," *The Guardian*, 16 February 2016.

34 Megan MacDonald, "Who Owns the Trauma Blanket? Tracing Kader Attia's
Ghost and the 'couverture de survie' in Transit," *Contemporary French and
Francophone Studies* 22, no. 2 (2018): 248.

35 See, for instance, the way in which reflectivity is used as part of the Drone
    Survival Guide, http://www.dronesurvivalguide.org/. For an example of
    stealth wear, see https://ahprojects.com/.

36 Kevin McHugh and Jennifer Kitson, "Thermal Sensations – Burning the
    Flesh of the World," *GeoHumanities* 4, no. 1 (2018): 157–77.

37 See, for instance, https://www.ukpreppersguide.co.uk/how-to-hide-your
    -infrared-thermal-image-or-body-heat-signiture/.

38 Elaine Yu, Steven Lee Myers, and Russell Goldman, "Hong Kong Protests:
    Over 1,000 Detained at a University, and a Warning From Beijing," *New York
    Times*, 18 November 2019.

39 Kara Thompson, *Blanket* (New York: Bloomsbury Publishing USA, 2018).

40 Alan Taylor, "Socialising in a Pandemic, Protected by Plastic," *The Atlantic*,
    26 May 2020, np.

41 Lucretius, *The Way Things Are: The De Rerum Natura of Titus Lucretius Carus*
    (Bloomington: Indiana University Press, 1968), 120.

42 Henri Bergson, *Matter and Memory* (New York, NY: Zone, 1991).

43 Engelmann and McCormack, "Elemental Worlds."

11 SEEDS

xan chacko

I wake up in Brisbane – *Meanjin* in Turrbal[1] – to news of the fire spreading in quantified terms. The human casualties in the tens. The buildings destroyed in the thousands. Acres burnt in the millions. Animals killed in billions. Since June 2019, bushfires have been burning across Australia at a rate, scale, and intensity that are unprecedented. Plants in the Australian bush are the medium that keeps these infernos burning, carrying the combustion to bewildering lengths and with devastating speeds. But these plants are, at the same time, invisible to the representational politics of loss. Represented purely in acreage of burn, they form the background upon which the devastation can be understood in human, economic, and moral terms yet plants themselves are missing in most accounts of the bushfires.

While great numbers of unique plant species will be lost forever in these fires – known now as the "Black Summer" – they do not illicit the affective gush for which the news media thirsts. In a rare article that specifically mentions the loss of plant species (alongside animal losses, of course), conservation ecologist Brendan Wintle is quoted in *The Guardian* saying, "You can lose the lot in one big fire."[2] Wintle continues, explaining that "If the timing is wrong, or the fire is too hot, you can also lose the *seed bank* and that's then another species on the extinction list."[3] Wintle refers to the bank of seed variety that makes up the population of a species in the wild. Even if individual specimens of the plant survive, a species is considered extinct if the bank is lost. While population banks of plants, such as the dark-bract banksia (*Banksia fuscobractea*) and the blue-top sun orchid (*Thelymitra cyanapicata*), are likely to have been destroyed in the recent bushfires, their loss does not garner international attention like that of animals or buildings. During a crisis, I have learned, plants are outside the limits of sympathy.

Large scale plant death is not limited to events such as the ongoing desiccation and burning of the Australian continent. Scientists, policy-makers, and social activists have been lamenting the extinction of plant life globally for at least a century, driven by anthropogenic devegetation for industry and agriculture. While the solutions have taken myriad forms, including reforestation and delimiting human encroachment, I focus here on ongoing research into one such techno-scientific endeav-our: cryogenic seed banking. A form of biological stockpiling, seed bank-ing emerged in the second half of the twentieth century as a potential solution to the crisis of loss of plant life diversity across the planet due to human encroachment. Seeds are placed in freezers to slow down their reproduction and ostensibly halt their decomposition until they are needed in the future. Taking seeds from their lived environment and literally putting them on ice serves to prolong, in our imagination, their life and renders them available to humans in the future. The logic of seed banking holds that human dependence on plants is so absolute that safeguarding plant futures saves humanity itself. At their simplest, seeds are elemental to more than just plant life, they contain within them the promise of our own species' survival. On closer inspection, seeds and the sciences that care for their long-term conservation expose vulnerabilities to thinking elementally. My provocation here is to think with the seeds to complicate the concept of the element.

Seeds are the fertilized embryos of plants that live today or have lived in the past. They also symbolize and embody the potential for future plants. Their construction as future plants is axiomatic to the practice of seed banking, and so the idea that seeds could die in the vault cannot be tolerated. The banks are liminal spaces that, at the same time, carry hope for the potential to sow future plants and also serve as grim reminders of the very vulnerability that necessitated the conservation of their seeds. Seeds in the bank are safe from the human and more-than-human rav-ages that plague everyday life. Seed banks thus collectivize human hope for survival because, however bleak the world outside may be, in the vault lies the potential for more life. My own work in the last decade has been to unravel the neat story of seed banks saving us from future star-vation, pulling at threads that support the edifice of seed banking. Some of these threads are rhetorical and others are embedded in the tech-no-scientific practices of the seed bank world. Exploring the work that goes into knowing seeds as elemental illustrates how willful ignorance is a requisite feature of maintaining the illusion of elementality. By observ-ing greenhouses and seed bank laboratories, I suggest, we can glimpse a more nuanced understanding of the precarity and interconnectedness of life, one that provides a much needed reality check of our efforts in plant conservation.

In their introduction to this volume, Neale, Addison, and Phan suggest that entry into an "Anthropogenic Table of Elements" requires entities such as seeds to fit at least one of the three defining characteristics of the elemental. In one common definition which the editors use, elements are identifiable discrete objects that can be discerned, studied, and combined. Most crucially, elements are indivisible without losing the basic characteristics of the substance, and seeds satisfy this sense of the elemental. They are functionally discrete in that they form the basis of identification of a whole plant species but also have the capacity to produce that plant from within themselves. You sow a seed and you reap a plant. In another sense, the elemental forms an environmental milieu or material substrate in which we find ourselves and other forms of life. This is embodied by the phrase "to be in one's element." Seeds bring their worlds with them. They contain all the necessary components to create a plant. Seeds contain both germ-able material as well as the initial food and environment for the fledgling plant. Seeds are substrate and substance combined. The third mode of elementality is metaphysical and makes a claim about the conditions-of-possibility of being. Something about life must be altered by the elemental in order to satisfy this criteria. By producing plants that will then go on to produce future plants, seeds also embody the condition of possibility of future life – of being and matter itself. Seen through this tripartite rubric, seeds are understood to be the basis of life and thus seemingly irrevocably elemental.

Any contributor to the Anthropocene epoch needs a time stamp or starting point in human history, something to point to and say, here, this is when humans intervened. For seeds, the moment could be at the advent of agriculture. Archaeobotanists argue that around 10,000 years ago, intrepid humans started intervening in the breeding of plants to secure stores of food and raw material for industries such as building and weaving.[4] Scientists understand this selection and amplification of certain varieties of plants over many generations to have caused an unnatural selection in the evolution of plants, separating the seeds of the plants we grow as crops from those of "wild" seeds. They mark this unwilding as the start of the anthropogenic influence over plants. Cultivation shares an etymological root with culture; they are primordially linked. In this sense, agriculture is at the basis of what it means to be human and thus is not only a symptom of the anthropos but a requirement of it. The effects of large-scale industrial agriculture as well as their synergistic chemical industries of fertilizers and pesticides have been noticed at the individual and global level.[5] Industrial agribusiness has come to emblematize the human imposition on an idealized imaginary of human-less nature.

Recognizing the corrupting influence of the anthropos over the imaginary of nature, and expecting that the endeavour of agribusiness will not abate, seed banks attempt to save what is left of "natural" life before it is lost forever. In many ways, this process of salvage mirrors the attempts of museums and anthropologists who attempt to capture "disappearing cultures" before they disappear altogether.[6] The subtle irony that it is the very success of modernity that causes the plants, people, and cultures to become extinct is not missed by the salvage processes. Conservators, collectors, and curators carry a guilt of being the cause of the loss, but at the same time they also evince the hope for survival for the "remnant objects" held within their collections. Remnant objects are entities that represent a broader kind but are also members of the kind and in some cases (especially those of species that are extinct) form the only examples of the kind.[7]

Iconic seed banks like the Svalbard Global Seed Vault (SGSV) in Norway summon the aesthetics of the sublime to create, what Neimanis, Åsberg and Hayes call a "climate imaginary."[8] Situated in the town of Longyearbyen, in the archipelago of Svalbard, at 78° north of the equator, the SGSV captures the curiosity of artists, philosophers, and scientists.[9] The SGSV opened in 2008 as the result of a tripartite agreement between Norway, which provide the space and security force to protect a cave in an Arctic mountain, the Crop Trust, who negotiate the agreements between the parties who send their seeds for storage, and the Nordic Genetic Resource Centre (NordGen), who manage the day-to-day care of the frozen collections. Taking on the remit of global conservation is as much an imaginary as the promise of long-term international cooperation that undergirds the maintenance and security of the collections. The SGSV's physical isolation and desolate location in the Arctic permafrost is the implicit first barrier of protection that seed banks in warmer accessible places lack. The territory of Svalbard, while technically part of Norway, is itself politically neutral as per an international agreement between the countries that share Arctic land. As the self-proclaimed guardians of future life, the SGSV has become endowed with a collective hope for the survival of humanity but is also inescapably always a symbol of the ruin of Nature.

Why does the onus of responsibility for safeguarding the future of life fall to industrialized nations of the Global North? Rather than totalizing and uniform, the conservation of plants through the banking of seeds provides another example of the variegated nature of the Anthropocene. Following geographer Erik Swyngedouw, I argue that some places have felt the apocalyptic effects of the Anthropocene more strongly than others, and that recent responses to the loss of plant life asymmetrically apportion blame to places in the Global South that are rich in biodiversity while, at the same time, granting responsibility for conservation to countries and institutions in the "biodiversity-poor," financially wealthy and industrialized Global North.[10] This paternalistic relation reiterates colonial forms of extraction and accumulation that anthropologists such as Michelle Murphy and Ann Laura Stoler remind me will continue to endure as long as they are inscribed within capitalism.[11] Recognizing the variance in the sensitivity towards anthropogenic change, Mario Blaser and Marisol de la Cadena propose that a pluriverse of many worlds exists within the world, many of which are not knowable by Western science.[12] While colonial legacies of extraction supply many of the banked seeds, nowadays, seed collecting is a political act undertaken after Prior and Informed Consent (PIC) and Access and Benefit Sharing (ABS) plans are painstakingly negotiated between seed banks and the people from whose country the seeds are collected.[13]

But seeds have never been truly elemental. What of the sunlight, water, microorganisms, insects, birds, soil, elevation, latitude, and human activities that comingle to spring forth living plants from seeds? Maria Puig de la Bellacasa, following Donna Haraway, reminds us that "nothing comes without its world."[14] This is what is so jarring about seed banking, which involves, first, a purposeful separation of the seed from its world and, second, its reinscription with meaning and value through the potential for future life that it holds. Seeds are rendered as "bare" potential – *zoë* as theorized by Giorgio Agamben – while all their prior trappings are killable as unwanted "debris."[15] The debris is incinerated for fear of pathogens that have come along for the ride from the home of the seed; every seed is precious and all debris is dangerous. At seed bank laboratories, curators spend hours cleaning, sorting, counting, testing, screening, and organizing seeds so that they are apprehensible to the system that gives them legitimacy. Using creative practices of care and violence, the scientists fashion worlds and meanings for the seeds in their new, and possibly forever, home.[16] Puig de la Bellacasa calls this kind of labour "a vital affective state, an ethical obligation and a practical labour."[17] As much as the seed bank world tries to control the liveliness of seeds, as living beings they neither recognize nor adhere to the borders created by humans to control their flows. Thus, seeds can never be strictly anthropogenic; they are always more-than-human. Seeds travel across international borders stuck to the wings of birds or inside their gastrointestinal systems. They practise interspecific sex and are also able to asexually reproduce themselves. Banana saplings, for example, can emerge from the underground corms of older banana trees.

The plant scientists with whom I engage also do not necessarily regard seeds as indivisible so that, in one respect, the elementary nature of seeds does not hold. The knowledge made in seed bank worlds are contingent on making seeds legible through the demarcating forces of classification, commodification and evaluation. Ascribing value to seeds through the valence of conservation and ecosystem services elevates them from living things to plant genetic material (or PGRs). Unlike seeds and plants, which have homes, people, and ecological companions, PGRs are defined through their future-looking capacity to produce more life or services to industry. The commodification of plants as PGRs to be used in the service of humanity – in a global sense – alienates the seeds from their local multi-species relations. This alienation allows the seed bank system to flourish as a saviour, without the messy imperial implications of extractions that accompany plants with traditional names and knowledges. What makes plants legible as PGRs is the isolation of traits like hairiness, drought resistance, or deliciousness, into DNA sequences that are cut and pastable in a synthetic biological imaginary.

Thinking of seeds as elemental serves very specific interests. Of the several hundred thousand known plant species, only some 120 are cultivated for human food. But just nine of these crops supply over 75 per cent of global plant-derived energy intake. Of these, only three – wheat, rice, and maize – account for more than 50 per cent of this intake. These are the seeds of agribusiness. These are the seeds of the Anthropocene. Plants grown as crops by humans that do not use seed for propagation include bananas, apples, all stone fruit, most tropical fruit, most vegetables that are grown from hybrids, and most flowers. Allowing these plants to reproduce sexually would randomize the traits for which they have been bred; their predictably shaped, coloured, and fragrant fruits, leaves, and stems that the marketplace of capitalism desires. Controlling plants' innate ability to reproduce for our material desires renders them sterile and frozen in evolutionary time. The seeds of plants that are not allowed to reproduce sexually are thus a thorn in the side of the conceptualization of seeds as elementary origins. Nothing starts from those seeds. And yet, seed banks save seeds and do so claiming the need to safeguard elementary futures.

The role of the seed bank seems to be less about practically seeding the future than it is about grounding the elemental view of seeds as starting points for future life. But what if the relation of plant, soil, air, insects and microbes is recalcitrant to commodification, multiplication, and co-optation into the biocapital of the vault? What happens when seeds are not enough? At the Millennium Seed Bank, the seed bank of the Royal Botanic Gardens in Kew, England, I spoke to Stacy, in 2016, who has been struggling to grow a set of plants from South Africa since 2006. The reasons to grow plants from seed banks' seed are myriad and could include needing to see the constituent parts of the plant for identification, or needing to multiply the seed collection if it was not sufficiently large to ensure its security in perpetuity. These plants in question refused to flower for over ten years. Despite Stacy's cunning attempts to coax them, they are refusing to bud. She suspected a symbiotic relationship between a bird or an insect in the home environment of the plant exists that she has not been able to approximate using tools that imitate different kinds of interspecific interactions. In a remarkable similarity to Charles Darwin's experiments in understanding the complex sensual relationship between wasps and orchids, Stacy's "involutionary momentum," while poetic and personal, has not affected the sexual dormancy of the plants.[18] Having tinkered, prodded and nudged in as many ways as she can muster, she was resigned to the idea that she did not have ultimate control over their reproductive desires. Being able to control by prediction the reproductive abilities of the seed is a primary concern in the matrix of valuation in seed banking.

The ability to claim with surety that the seeds that have been cared for and banked will be able to reproduce themselves is a basic expectation of seeds in the bank. However, listening to Stacy's concerns reopens this expectation as a goal and desire rather than a given. Stacy asked me, "Why should we expect these plants to thrive without their ecological companions?" This question cannot be answered within the seed bank's logic because it threatens to undermine the very nature of the project. There are, however, significant consequences to not being able to control the reproduction of the seeds in the bank. If Stacy is not able to coax a plant into producing flowers and fruits, its identity could remain unsettled. The seeds of this plants are saved but, recalling the criteria for elementality, what value could they have as elementary indicators without a name for their plant origins? What value could they have if they are unable to grow to full maturity or reproduce, meaning they are out of their element? And, most importantly, what value could they have if they do not act as representatives of their kind in their capacity to repopulate the world? In a speculative imaginary, if these seeds were returned to their home environment, there is a chance that severed ties could be redrawn. Ever the feminist killjoy, let me ruin the nostalgia of this suggestion with a reminder that the term used by scientists for bringing seeds back to their country of origin is "repatriation." Seeds are considered patrimony. Seeds are property inherited from one's male ancestor.

Could, instead, a capacious theory of the seed help clarify what it means to be elemental? The requirement of thinking from the seed involves an implosion of the idea of elementarity. Picture, if you have not tried it, sprouting seeds yourself. The root and shoot emerge from the seed itself and thus contain both the plant and the medium for growth, both substance and substrate, figure and ground. Seeds provide an inversion to the modular dimension of the elemental, demanding that we complicate the concept. In trying to pare botanical life down to seeds, I see that seeds are always in excess and yet at the same time are not enough. Seeds' practices open up the concept of the elemental to reveal that thick interconnectedness is unavoidable in any attempts to sustain life. As Stacy's project and my home gardening attempts attest, sprouting the seeds is not enough. Keeping them alive, cajoling them to flower and fruit, and caring for their needs is a lot tougher to do predictably. Stacy's frustrations reveal an inadequacy to know and master plants that haunts the sciences so dedicated to controlling life. Plants are in excess of the anthropos. In fact, humans only live because of the elemental alchemy of photosynthesis. Geographers Nigel Clark and Kathryn Yusoff remind us that the fossil fuels, whose burning in earnest marks the start of the Anthropocene for many, are but plant bodies that lived and died hundreds of millions of years ago.[19] As I helplessly watch landscape fires burn living and dead plants in California and Australia, I find solace in Clark and Yusoff's provocation to "queer" fire as I've been queering plants. Clark and Yusoff helpfully connect the queerness of plants to the erotic energies of fire by pointing out that the medium that burns in these and other fires is the product of the procreative proclivities of plants, which adhere neither to heteronormativity nor the species boundary. Keeping plant and seed stories alive through their retelling draws me into relation with them at a time when their worlds are being destroyed on purpose and without intention. Every seed story is partial and incomplete because to tell the whole story would be a "god trick" tantamount to knowing and capturing life itself.[20]

In Meanjin, visual artist Sophie Munns provides a venue to engage with seeds in her SeedArtLab. In 2010, Munns launched the "Homage to the

Seed" project as part of a residency at the Brisbane Botanic Gardens. Since then, she has travelled to several seed banks, gardens, and community organizations to understand the science of conservation at a time of shifting climate conditions. Munns has shared her findings through a unique visual language that abstracts the seeds from their clinical surroundings in the vault without evoking a nostalgia for their home environments. At her studio, Munns hosts workshops for small groups of all ages in order to build relationships between people and the seeds that live around them. In 2019, I joined Munns for an afternoon SeedArtLab workshop held specifically to raise funds for fire-affected communities in Australia. This is the way that Munns knows to cope with the devastation surrounding us: by facilitating engagement with seeds through intimate drawing groups.

I find myself sitting behind a table filled with seeds from Munns's personal collection. I am invited to draw *anything* I want to using twigs of assorted sizes and textures as brushes, held in a circular stand, and a pot of thick, sticky, black ink. The experience of confronting the blank piece of paper forced me to reckon with how little confidence I have in my ability to draw anything resembling the textures and forms of the seeds I have in front of me. Having encountered, appreciated, and worked with seeds during the last ten years of research at seed banks makes my inability to depict them in this moment all the more terrifying. Munns nudges me to start, make an impression, and then move on, freeing me to let go of the idea that perfection of representation was at all possible or desired. The process Munns describes required a rejection of mastery, the kind of speculative feminist practice that requires humility and is accepting of limitations. With tea and cake, the afternoon workshop continued, and I let go of the expectations that I had built up for myself and participated in the creation of a kind of climate imaginary (see figure 11.1).

Talking and laughing with the other participants in the workshop brought up memories for each of us, memories of relating to particular seeds through our childhoods. For Munns, this is the possibility that art opens up, not against but alongside the institution of seed banking. Her work in meaning making through embodied practice drew us together as knowers of seed bodies, histories, and futures. Since relating is so fundamental to the experience of the anthropos, the elemental simplicity of this exercise has the potential to create an unbounded experience with seeds, an appreciation of their weird and wonderful bodies without the identification, valuation, and valorization in service of humanity. Being limited in my ways of expressing seeds through twig and ink drew me closer to the scientists like Stacy who are trying to imagine, like I am, a

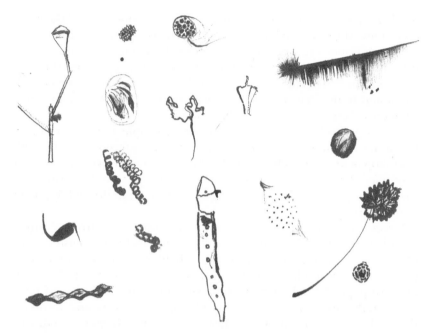

Figure 11.1.   Seed drawing at SeedArtLab, 15 December 2019 (credit: Xan Chacko)

concept for seed futurity. Confronting these seeds and failing to capture their totality in neither banking, drawing, nor breeding, I can only hope that some of them are still out there. Despite the fire. Despite the warming climate. Despite me.

NOTES

1   Turrbal is the Aboriginal Australian language and also the name of the people who are the traditional custodians of the region that comprises present day Brisbane, Australia. The place name Meanjin is also interchangeably used to refer to the people.
2   Graham Readfearn, "'Silent Death': Australia's Bushfires Push Countless Species to Extinction," *The Guardian*, 4 January 2020.
3   Emphasis mine.
4   BrieAnna S. Langlie, Natalie G. Mueller, Robert N. Spengler, and Gayle J. Fritz, "Agricultural Origins from the Ground Up: Archaeological Approaches to Plant Domestication," *American Journal of Botany* 101, no. 10 (2014): 1601–17.

5   Rachel Carson, *Silent Spring* (New York, NY: Houghton Mifflin Harcourt, 1962).

6   Rebecca Lemov, *Database of Dreams: The Lost Quest to Catalog Humanity* (New Haven: Yale University Press, 2015).

7   James R. Griesemer, "Modeling in the Museum: On the Role of Remnant Models in the Work of Joseph Grinnell," *Biology and Philosophy* 5, no. 1 (1990): 3–36.

8   Astrida Neimanis, Cecilia Åsberg, and Suzi Hayes, "Post-Humanist Imaginaries," in *Research Handbook on Climate Governance*, eds. Karin Bäckstrand and Eva Lövbrand (Northampton: Edward Elgar Publishing, 2015), 480–90, 484.

9   Xan S. Chacko, "Calm Yourself, the World Is Ending," *Sydney Environment Institute Opinion*, 6 May 2019, http://sydney.edu.au/environment-institute/blog/calm-world-ending/.

10 Eric Swyngedouw, "Apocalypse Now! Fear and Doomsday Pleasures," *Capitalism Nature Socialism* 24, no. 1 (2013): 9–18.

11 Michelle Murphy, *The Economization of Life* (Durham, NC: Duke University Press, 2017); Ann Laura Stoler, *Duress: Imperial Durabilities in Our Times* (Durham, NC: Duke University Press, 2016).

12 Mario Blaser and Marisol de la Cadena, "Pluriverse: Proposals for a World of Many Worlds," in *A World of Many Worlds*, eds. Mario Blaser and Marisol de la Cadena (Durham, NC: Duke University Press, 2018).

13 Xan S. Chacko, "Digging Up Colonial Roots: The Less-Known Origins of the Millennium Seed Bank Partnership," *Catalyst: Feminism, Theory, Technoscience* 5, no. 2 (2019): 1–9.

14 Maria Puig de la Bellacasa, "'Nothing Comes without Its World': Thinking with Care," *The Sociological Review* 60 (2012): 197.

15 Giorgio Agamben, *Homo Sacer: Sovereign Power and Bare Life* (Stanford: Stanford University Press, 1995).

16 Xan S. Chacko, "Creative Practices of Care: The Subjectivity, Agency, and Affective Labor of Preparing Seeds for Long-Term Banking," *Culture, Agriculture, Food and Environment* 4, no. 2 (2019): 97–106.

17 de la Bellacasa, "Nothing Comes Without Its World," 197.

18 Carla Hustak and Natasha Myers, "Involutionary Momentum: Affective Ecologies and the Sciences of Plant/Insect Encounters," *Differences* 23, no. 3 (2012): 74–118.

19 Nigel Clark and Kathryn Yusoff, "Queer Fire: Ecology, Combustion and Pyrosexual Desire," *Feminist Review* 118 (2018): 7–24.

20 Donna J. Haraway, "Situated Knowledges: The Science Question in Feminism and the Privilege of Partial Perspectives," *Feminist Studies* 14, no. 3 (1988): 581.

12 SPERM

janelle lamoreaux
ayo wahlberg

In global environmental movements, sperm has yet to achieve the iconic status of collapsing ice shelves, starved polar bears, and overflowing garbage mountains. Yet andrologists and sperm bank coordinators increasingly suggest that one can measure the health of the environment through an assessment of sperm. "To figure out whether an ecosystem is stable or not, all you have to do is test the sperm ... [if] the environment is bad, sperm become ugly," said Director Li Zheng of Shanghai's Sperm Bank in 2013, at the height of China's "sperm bank emergency" (*jingzi ku gaoji*) when demand for donor sperm far surpassed the national supply.[1] What is "ugly sperm"? In this chapter, we suggest that thinking about sperm elementally helps us to not only answer this question, but also to trace the intricate and inseparable linkages between planetary liveability and (more-than) human reproduce-ability in a time of environmental crisis.

Amid intensifying global exposures, human sperm has become yet another of the Anthropocene's sentinels – a living being or technical device that warns of catastrophe to come.[2] While geologists, marine biologists, and atmospheric scientists warn of a bleak planetary future that is compromising the earth's biocapacity and thereby liveability, toxicologists, andrologists, embryologists, and other reproductive and developmental biologists have begun warning of a bleak reproductive future as human "reproduce-abilities" are compromised by these same anthropogenic processes. As endocrinologist Niels Erik Skaakebaek suggests, "The low fertility rates we have had for decades by now, are an existential threat ... to all industrialized countries."[3] Alongside melting glaciers, disappearing rainforests, transforming fish and threatened insects, declining and damaged sperm have become a warning signal. Sperm are an indicator of humanity's unsustainable decline, and a substance through which extensive relationships between these "elemental" cells – some of the most basic building blocks of a potential person – and their toxic surroundings are understood.

We begin our chapter by chronicling a scientific debate that was kick-started in 1992 following the publication of a review article in the *British Medical Journal* by andrologist Elizabeth Carlsen and colleagues. This meta-analysis of sperm science from around the globe concluded that "semen quality ... has declined appreciably during 1938–90 ... Such remarkable changes in semen quality ... over a relatively short period [are] more probably due to environmental rather than genetic factors."[4] We show how scientists in China have subsequently engaged with this particular debate, as worries about the detrimental effects of industrial pollution on fertility have circulated and shaped research agendas and socio-technical imaginaries in recent decades. We then move on to discuss the intergenerational dimensions of sperm research, showing how questions about individualized fertility in the present quickly expand to a reckoning with humanity's future existence and reproduce-ability. We conclude by suggesting that sperm has become elemental in the "age of the human," raising profound questions about reproductive futures.

Ever since Carlsen and colleagues published their 1992 review article, a global debate has ensued about whether (and if so to what extent) sperm quality is falling in countries around the world. The semen that sperm cells swim in is made up of fructose, amino acids, flavins, enzymes, zinc and more, while the sperm cells themselves, like other living cells, are composed primarily of carbon, hydrogen, oxygen and nitrogen. Sperm cells are haploid germ cells, providing half of the 46 chromosomes found in human somatic cells when they fuse with an egg cell containing the remaining half during fertilization. Technological advances have resulted in the development of what Lisa Jean Moore and Matthew Schmidt have called techno-semen, meaning "semen that has been technically manipulated in laboratory environments, has been collected from men who have been selectively screened for genetic and social characteristics, and has been rhetorically constructed by semen banks' industrial marketing strategies."[5] Despite this, assessing the quality of sperm in many sperm banks remains, in large part, a scientific craft.[6]

Semen analyses often rely on the skilled eye of the andrologist, who sits perched at a microscope counting the number of visible sperm cells against a gridded backdrop, evaluating the liveliness of the sperm cells in terms of how fast they are moving under the lens of the microscope and judging the morphology, if not aesthetic beauty, of individual sperm cells. This craft notwithstanding, such semen tests are (like much in high-throughput biological work) increasingly being replaced by computers with software algorithms that have been trained to assess quality. While the exact biological mechanisms remain unclear, reproductive scientists have observed a link between sperm quality and fertility. This has led to the development of third-party sperm banks on the one hand (initially for men who are living with azoospermia, unable to produce their own sperm) and, on the other hand, specialized techniques aimed at extracting even single sperm cells from men's testes (to be used in intracytoplasmic sperm injection as a means of fertilization during fertility treatment).

With infertility rates apparently on the rise throughout the world and use of reproductive technologies certainly increasing, if not booming in China and elsewhere, the question of whether or not sperm quality is falling has become a matter of reproductive politics within individual nation states as well as globally. After a 1995 study from France seemed to support Carlsen and colleagues' conclusion that semen quality was indeed declining,[7] a series of peer critiques were put forth. Taking issue with the two studies' methodologies, they argued for the impossibility of comparing different studies given that practices of assessing sperm quality were subjective to such an extent that it was almost impossible to retroactively quality-control the sperm quality assessments that undergirded prior studies. Richard Sherins of the Genetics and IVF Institute in Fairfax Virginia was emphatic:

> Several studies over the past 20 years have suggested that the quality of semen is declining in industrialized countries throughout the world, arousing concern about male fertility in the future ... It would seem prudent, however, to ask whether sperm counts and male fertility are in fact declining. Unfortunately, the literature on this subject is not based on prospective studies designed to assess changes in semen quality and male fertility among members of the general population ... Rates of infertility have remained constant during the past three decades (at 8 to 11 per cent), and male infertility has accounted for approximately one third of cases.[8]

These critiques notwithstanding, the suggestion of population-level declines in sperm quality was now circulating, as was the suggestion that this decline might be linked to the industrial chemicals that men were exposed to both prenatally and throughout their life courses. Indeed, these claims were popularized in the 1996 best-selling US book *Our Stolen Future: Are We Threatening Our Fertility, Intelligence and Survival?*, by Theo Colborn, Dianne Dumanoski, and John Peterson Myers. In continuing the environmental agenda that American activist Rachel Carson had set with her 1962 book *Silent Spring*, in which she warned of toxic chemicals becoming "lodged in all the fatty tissues of the body."[9] It was the book by Colborn and colleagues that brought wide attention to the negative effects of "endocrine disruptors" (to which we will soon return). One of the exemplar cases they used was that of falling sperm quality, and, as they reflected in the book's second edition: "No aspect of the extensive scientific evidence considered in *Our Stolen Future* commanded more media attention than the controversial reports over the past four years of declining sperm counts."[10]

Scientific debates about (apparently) falling sperm quality were also taking place in China in the late 1990s, just as the effects of Deng Xiaoping's Four Modernizations program in agriculture, industry, defence and science and technology had resulted in mass market-oriented industrialization and urbanization. In 1999, researchers from the National Population and Family Planning Commission in Beijing analysed data on 11,726 semen assays collected in 39 cities and counties from 1981 to 1996, reporting that sperm motility had decreased from 75.11 per cent to 67.27 per cent and that the percentage of sperm with normal morphology had fallen from 85.02 per cent to 77.89 per cent.[11] In their conclusion, Zhang and colleagues made the point "that although Chinese sperm quality is [currently] higher, it declines significantly faster than that of western countries at the same period. It is possible that the decline of sperm quality is due to environmental quality."[12] While the same methodological issues that critics of Carlsen's study had raised were just as relevant in China,[13] this study would nevertheless play an important role in cementing within national debates a posited link between environment and apparent falls in sperm quality. Moreover, it was around this time, in the late 1990s, that a number of improvised sperm banks emerged throughout the country, from Qingdao to Chongqing. As Wang Yixing of Renji Hospital in Shanghai explained in 2001, "sperm banks seemed to be popping up all around the country overnight, and some of them can't ensure the sperm source and quality."[14]

Given China's strict regulation of family planning, such a state of affairs was not tenable and indeed, in 2003, the country's first regulations on Assisted Reproductive Technologies came into force requiring all fertility clinics and sperm banks to adhere to strict licensing requirements. At the same time, evidently rising demand for and use of sperm banks throughout China were seen by some fertility experts as cause for concern. Professor Liu Dalin from Shanghai University echoed the concerns that Theo Colborn and others had raised a few years earlier, warning in a 2008 interview, "do not let Man become an endangered animal."[15] His concerns seemed to be corroborated as a number of cohort studies from researchers based in sperm banks throughout China began reporting aggregate decreases in sperm concentration, the percentage of sperm with normal morphology and total sperm count among would-be donors.[16] Moreover, beginning in the 2000s, media reports began referring to national permutations of a global "sperm crisis" (*jingzi weiji*) that followed from findings of generalized semen decreases over time, as well as a national emergency in China's sperm banks (*jingzi ku gaoji*). These nationally governed facilities were experiencing shortages and a campaign to find suitable donors was soon reinvigorated. Altruistic donation was encouraged for a variety of reasons. Interestingly, potential sperm donors interviewed in the early 2010s at China's largest and oldest sperm bank in Changsha, capital of Hunan province, pointed to the "serious problem of the environment" as one of their motivations for donating.[17]

To be sure, the question of whether or not population-level sperm quality is falling in China or elsewhere remains contested. Yet, what three decades of mass-mediated scientific debate have ensured is that falling fertility has come to be viewed as a kind of harbinger. Is sperm quality changing? Are sperm counts falling? Is male infertility rising? And, if so, are industrial chemicals and pollution to blame? The jury may be out, but these matters of concerns now mobilize scientists, governments, environmental organizations, and infertile patients alike, not least in China. As the headline in a December 2013 article from the *South China Morning Post* reads, "Smog crisis in China leads to increased research into effect of pollution on fertility – Beijing's funding for research into how chronic pollution is affecting childbearing triples in last five years."[18]

Why does a decline in semen quality mobilize so many? What is it about the possibility of falling male fertility that attracts the attention of scientists, governments, and environmental organizations? Certainly, some of sperm's provocative potential can be attributed to a perceived threat to male virility, as political scientist Cynthia Daniels suggests. Daniels argues that the risks of environmental exposure to men's reproductive health were long downplayed and remain contested due to stereotypes of masculine strength.[19] Likewise, in China, Everett Zhang has shown how the medical sub-speciality of *nanke* ("Men's Medicine") emerged in the 1980s primarily as a response to what was seen as an impotence epidemic rather than a concern for male infertility.[20] But, in research on the relationship between declining human reproductive health and changing environments conducted by andrologists, toxicologists, and other scientists today, masculine bodies and bodily substances often take centre stage, at its most extreme leading to scientific debate on the complete degeneration of the Y chromosome in evolutionary time.[21] Even in popular representations of an apocalyptic infertile future, as in novels such as Margaret Atwood's *The Handmaids Tale* (1985) and Zhang Xian Liang's *1.6 Hundred Million* (2012), men are represented as uniquely vulnerable to environmental destruction. Such gendered emphasis is particularly present in the science and popular representation of endocrine disrupting chemicals (EDCs).

EDCs are chemicals classed by their ability to negatively affect humans and other animals through the endocrine system, producing a wide range of problems with developmental, reproductive, neurological, and immune health.[22] Some endocrine disruptors are naturally occurring, such as phytoestrogens, but many more are man-made. Over 1,000 chemicals are now classified as EDCs, including DDT (Dichlorodiphenyltrichloroethane) and PCBs (Polychlorinated biphenyls), as well as many flame retardants and pesticides. EDCs are often described as "mimics" of naturally occurring hormones, which interfere, disrupt, or perturb hormonal systems through estrogenic effects.[23] Research on EDCs often focuses on their reproductive and developmental health effects, stressing their feminizing and gender-altering qualities. Activist campaigns, particularly those based in Europe and the United States,[24] gender the problem even further, focusing on the effects of what are sometimes referred to as "gender-bending" chemicals or an "an assault on the male."[25] As a rhetorical strategy, the depiction of sperm emasculated by estrogenic environments plays on gendered stereotypes long-affixed to reproductive cells.[26]

Another dimension of EDC research that is particularly relevant to sperm is a focus on potential intergenerational effects. Since sperm cells are germ cells, environmentally induced DNA damage in sperm has been flagged as a potential conduit of health problems in future generations.[27] The question, increasingly being asked by journalists and scientists, is if sperm is damaged by the pollutants, pesticides, and plastics of the day, might such damage be carried onto future generations? Historian Scott Frickel argues that interest in damaged sperm DNA is not only about specific effects on future generations, but also a more general concern about the integrity of the human gene pool.[28] Frickel points to a "subtle rhetorical symbiosis" between ideas of DNA damage and eugenics, occurring in a century where the gene takes on an increasingly important role in the popular and scientific imagination.[29] Following from this, it seems that widespread attention to the health of human sperm in the Anthropocene goes far beyond concern about individual infertility and virility, to evoke anxiety about the twinned environmental decline and genetic decline of a people or a nation more broadly.

Such rhetorical synergy between environmentally-induced sperm decline and a broader decline in population quality (*renkou suzhi*) became apparent in 2013, when the "ugly sperm" news broke in China. Although sperm bank shortages had been regularly reported in China, due to a number of factors including strict regulations on donation and donor characteristics, comments from the director of a Shanghai-based sperm bank directly connected sperm decline to environmental factors and resonated with many. Reactions to this expert opinion on a sperm-environment correlation in print and social media were wide ranging. Some wondered if the cost of the nation's economic development was the sacrifice of the next generation, while others expressed concern about the increased risk of dying without male heirs or offspring (*duanzijuesun*), and the potential threat of ugly sperm to the Chinese nation and its people (*zhonghua minzu*).[30] In such instances, sperm – even at its microscopic scale – opens up anxious reflection on intergenerational elements and even on the extinction of a lineage, an ethnicity, or a nation.

In a context that has stressed the need for an increase in "population quality" since the late 1970s, and has more recently been concerned with growing a population previously restricted through birth planning policies, the stakes of sperm threatened by a deteriorating environment could not be higher. Following four decades of birth quotas, pregnancy certificates, sterilizations, and abortions, family planning officials in China are now faced with the question of how to encourage births in a time of falling fertility, whether through broader access to assisted reproductive technologies, work-life balance initiatives, or more draconian impositions of deposits (to be returned after the birth of a second child). Amid such an ongoing reconfiguration of China's reproductive complex, the effects of environmental exposure on fertility has become one more factor to attend to if the new characterization of China as a low fertility country is to be changed, and Chinese families reproductive behaviour – previously considered "problematic" – restored. At the same time, however, given the brutal legacies of China's one-child policy (1980–2015), especially for women, such official encouragement or requirement of birth is rendered unbearably ironic.

Apocalyptic thinking about the sperm crisis can problematically reinforce hetero- and bio-normative ideas of reproduction and gender, currently enshrined in China's marriage and maternal and child healthcare laws, as well as their nationalist and eugenic undertones. But it can also open ideas about the intrinsic relationship of cells to their surroundings. Attention to declining sperm as a measure of masculinity, a reflection of the environment surrounding it, and an intergenerational conduit of potential damage in future generations only seems to be increasing. But this attention is not, and may never have been, simply about an environmental threat to individual men's fertility and virility. Measures of semen quality have become a barometer for the reproduce-ability of a specific mode of economic, social, and intergenerational organization in the present and the potential future. If the very "building blocks of life" in the form of elemental sperm cells are at risk, a nation like China – that has long linked economic development and national strength to strict control of reproductive behaviour – is compelled to take fertility experts' warnings about (potential) detrimental effects of exposure seriously as a national matter of concern.

Measurements of tree rings, riverbed deposits, and mother's milk have all been used to gauge the extent to which human activity has irrevocably changed the environment, and toxic exposure changed the human. Like these elemental traces, sperm are now thought to be material instanti-ations and symbolic figures that speak beyond individual bodies to the health of what stands outside them. From sperm counts and measures of motility and morphology, to sperm DNA tests that describe the extent of chromosomal damage wrought by EDCs, sperm tests have become more than analyses of individual semen samples. They have become a mode of analysing environmental destruction, and its gendered, genetic, and intergenerational costs. As a sentinel, "ugly sperm" brings attention to the way bodies and bodily substances carry within them a record of environmental exposure. Sperm today are understood as not just an el-emental part of humans in the present and future, but also sentinels of humans' negative influence on the reproduce-ability of planetary life in an anthropogenic age.

NOTES

1 Janelle Lamoreaux, "Making a Case for Reducing Pollution in China, or the Case of the Ugly Sperm," *Somatosphere* (2015); Ayo Wahlberg, *Good Quality: The Routinization of Sperm Banking in China* (Oakland, CA: University of California Press, 2018).

2 Frederic Keck and Andrew Lakoff, "Figures of Warning," *Limn* 3 (2013).

3 Cited in Lars Igum Rasmussen, "Det store forventede babyboom er fuldkommen udeblevet," *Politiken*, 22 February 2020.

4 Elisabeth Carlsen, Aleksander Giwercman, Niels Keiding, and Niels E. Skakke-bæk, "Evidence for Decreasing Quality of Semen during Past 50 Years," *British Medical Journal* 305, no. 6854 (1992): 609–13.

5 Lisa J. Moore and Matthew A. Schmidt, "On the Construction of Male Differences: Marketing Variations in Technosemen," *Men and Masculinities* 1, no. 4 (2018): 331–51.

6 Mianna Meskus, *Craft in Biomedical Research: The iPS Cell Technology and the Future of Stem Cell Science* (London: Palgrave, 2018).

7 Jacques Auger, Jean Marie Kunstmann, Françoise Czyglik, and Pierre Jouannet, "Decline in Semen Quality among Fertile Men in Paris during the Past 20 Years," *New England Journal of Medicine* 332, no. 5 (1995): 281–5.

8 Richard J. Sherins, "Are Semen Quality and Male Fertility Changing?" *New England Journal of Medicine* 332, no. 5 (1995): 327–8.

9 Rachel Carson, *Silent Spring* (New York: Houghton Mifflin Harcourt, 1962); Ayo Wahlberg, "Exposed Biologies and the Banking of Reproductive Vitality in China," *Science, Technology and Society* 23, no. 2 (2018): 307–23.

10 Theo Colborn, Dianne Dumanoski, and John Peterson Myers, *Our Stolen Future: Are We Threatening Our Fertility, Intelligence and Survival? – A Scientific Detective Story* (New York: Penguin, 1996).

11 S.C. Zhang, H.Y. Wang, and J.D. Wang, "Analysis of Change in Sperm Quality of Chinese Fertile Men during 1981–1996 [in Chinese]," *Reproduction & Contraception* 10, no. 1 (1999): 33–9.

12 Zhang et al., "Analysis of Changing Sperm Quality," 39.

13 *Sina News*, "Sperm Quality Is Declining and Male Infertility Is Rising," 25 September 2008.

14 Sunny Hu, "Where There's Sperm, There's Demand," *Shanghai Star*, 3 May 2001; see also Wahlberg, *Good Quality*, 2018.

15 *Sina News*, "Sperm Quality Is Declining," 2008.

16 Chuan Huang, Baishun Li, Kongrong Xu, Dan Liu, Jing Hu, Yang Yang, Hong Chuan Nie, Liqing Fan, and Wenbing Zhu, "Decline in Semen Quality among 30,636 Young Chinese Men from 2001 to 2015," *Fertility and Sterility* 107, no. 1 (2017): 83–8; M. Jiang, X. Chen, H. Yue, W. Xu, L. Lin, Y. Wu, and B. Liu, "Semen Quality Evaluation in a Cohort of 28,213 Adult Males from Sichuan Area of South-West China," *Andrologia* 46, no. 8 (2014): 842–7; Meng Rao, Tian-Qing Meng, Si-Heng Hu, Huang-Tao Guan, Qin-Yu Wei, Wei Xia, Chang-Hong Zhu, and Cheng-Liang Xiong, "Evaluation of Semen Quality in 1,808 University Students, from Wuhan, Central China," *Asian Journal of Andrology* 17, no. 1 (2015): 111; Li Wang, Lin Zhang, Xiao-Hui Song, Hao-Bo Zhang, Cheng-Yan Xu, and Zi-Jiang Chen, "Decline of Semen Quality among Chinese Sperm Bank Donors within 7 Years (2008–2014)," *Asian Journal of Andrology* 19, no. 5 (2017): 521.

17 Wahlberg, *Good Quality*, 77–99.

18 *South China Morning Post*, "Smog Crisis in China Leads to Increased Research into Effect of Pollution on Fertility," 11 December 2013.

19 Cynthia R. Daniels, *Exposing Men: The Science and Politics of Male Reproduction: The Science and Politics of Male Reproduction* (Oxford: Oxford University Press, 2006).

20 Everett Yuehong Zhang, *The Impotence Epidemic: Men's Medicine and Sexual Desire in Contemporary China* (Durham: Duke University Press, 2015).

21 Sharyn Davies and Samuel Taylor-Alexander, "Temporal Orders and Y Chromosome Futures: Of Mice, Monkeys, and Men," *Catalyst: Feminism, Theory, Technoscience* 5, no. 1 (2019): 1–18.

22 Malin Ah-King and Eva Hayward, "Toxic Sexes: Perverting Pollution and Queering Hormone Disruption," *O-Zone: A Journal of Object-Oriented Studies*, no. 1 (2013): 1–14.

23 Sheldon Krimsky, *Hormonal Chaos: The Scientific and Social Origins of the Environmental Endocrine Hypothesis* (Baltimore: Johns Hopkins University Press, 2000); Celia Roberts, *Messengers of Sex: Hormones, Biomedicine and Feminism* (Cambridge: Cambridge University Press, 2007).

24 Janelle Lamoreaux, "'Swimming in Poison': Reimagining Endocrine Disruption through China's Environmental Hormones," *Cross-Currents: East Asian History and Culture Review* 8, no. 1 (2019): 195–223.

25 BBC Productions, "The Estrogen Effect: Assault on the Male" (1993).

26 Emily Martin, "The Egg and the Sperm: How Science Has Constructed a Romance Based on Stereotypical Male-Female Roles," *Signs* 16, no. 3 (1991): 485–501.

27 Melissa Perry, "Chemically Induced DNA Damage and Sperm and Oocyte Repair Machinery: The Story Gets More Interesting," *Asian Journal of Andrology* 17 (2015): 1–2

28 Scott Frickel, *Chemical Consequences: Environmental Mutagens, Scientist Activism, and the Rise of Genetic Toxicology* (New Brunswick, New Jersey, London: Rutgers University Press, 2004).

29 Frickel, *Chemical Consequences*, 104. See also Evelyn Fox Keller, *Century of the Gene* (Cambridge: Harvard University Press, 2002).

30 Janelle Lamoreaux, *Infertile Environments: Epigenetic Toxicology and the Reproductive Health of Chinese Men* (Durham: Duke University Press, forthcoming).

13 STRONTIUM

brad bolman

The young women of the Pleasant Grove 4-H Club enjoyed most aspects of their tour of the University of California's agricultural school at Davis, but their foremost pleasure was the dogs. Rose Ann Sharpless, Dottie Ormsby, Diane Burke, Laurel Rodriguez, Carolyn Schellhous, and her sister Alice, Nickie Schultz, Mary Ellen Rosenberg, and Joyce Bettes walked the grounds of a project known affectionately among UC administrators as "Project Hot Dog," while Dottie's parents and Mrs. Schellhous, who went by June, looked on. J.W. Kendrick, the new chair of Medicine, Surgery and Clinics in the Veterinary School, led the girls through the veterinary clinic, introducing them to X-ray machines and the operating rooms. In 1961, the trip was unexceptional, as Sacramento's nuclear families often packed themselves into vehicles on the weekend and set out to see the beagles. Yet all that remains of the Pleasant Grove 4-H Club's trip and that of so many other families is a short note, a trace, in Marysville, California's Appeal-Democrat from 11 April 1961, preserved online in a database of regional papers, a graveyard for America's dwindling local press.[1]

One could fairly say that the historian of science – whether of physics, chemistry, or biology – is a collector and analyst of traces, writing "a history of objectivity," in the words of Hans-Jörg Rheinberger, from these material-semiotic remainders.[2] So too is the animal historian, who must "look for the traces of the animal, animals, or animality in even the most human of texts," as Etienne Benson writes.[3] The work of both scholars is of a piece with what we might call a broader "elemental" imagination: the assumption of an ultimate identity between past and present that following the traces – archival and chemical, animal and human – allows us to locate. In what follows, then, I reflect on the multivocality of the "trace," a concept that threads across the persistence of nuclear containment, the ongoing challenges of environmental safety, the survival of scientific records – and even the notion of truth itself. We might begin tracing these traces from the resolutely chemical.

Few families in the Davis area now remember the beagles. During a lull in my visit to the site in 2016, after passing through quiet halls that once thrilled Nickie and wowed Joyce – or perhaps the other way around – I was shown an empty, square alley. The small patch of rain-soaked land (see figure 13.1) had been remediated years earlier by the American government, but they left behind larger landfills of contaminated materials, deemed the responsibility of a cash-strapped UC system. In October 2018, $14 million was finally allocated to reduce the residual waste elements to a trace, the penultimate step in a decades-long effort to cleanse grounds once occupied by hundreds of strontium- and radium-fed beagles.[4] Only a few aging buildings remain standing nearby.

The final canines disappeared in the 1990s, when the survivors of Davis's large breeding colony were variously adopted out by the Marin Humane Society and repurposed for newly popular research on Alzheimer's. Most, however, vanished in the 1980s as funding and interest in this formerly grand Cold War project dried up: first the dogs passed away, then their radioactive carcasses were transported by the Department of Energy to the grounds of the Hanford Site in eastern Washington for permanent storage.[5] Their bodies were once valuable specimens, keys to understanding the risks of strontium-90, a bone-seeking isotope released in nuclear explosions. But as Bern Shanks, former director of environmental health and safety at Davis, noted in 1990, "They're no longer cute little dogs; they're just a radioactive waste problem."[6]

The problem went beyond their physical bodies: tons of radioactive canine fecal materials were also sent to Hanford in the final decades of the twentieth century. An initial research oversight – that dogs fed radioactive materials would keep in only about 2 per cent of the irradiation – had driven engineer Edward Edgerley Jr. to develop an Imhoff tank system for "high-order decontamination using cation exchange resins" for his PhD in sanitary engineering at Berkeley. Former lab director Marvin Goldman gave Edgerley's device a more candid moniker: the "Radioactive Poop Machine," yet even the RPM failed to completely eliminate byproducts of the beagle "waste stream." As Zygmunt Bauman argues, one condition of modernity is the impossibility of complete disappearance. In our world, "no waste can be disposed of radically and completely" –

Figure 13.1. Remediated site at University of California, Davis, 2016 (credit: Brad Bolman)

not even dog shit. Waste is itself a trace, one that can only be hidden or recycled: shit spun into gold.[7]

Opened in the early 1950s and planned originally to study the safety of pilots in nuclear-powered aircraft, the Davis Lab's pivot to studying strontium reflected the element's increasing prominence as a possible product of fallout from nuclear weapons testing. Strontium, a yellowish metal, holds the ambiguous honour of being the periodic table's only element named for a city in the United Kingdom. The chemist Thomas Charles Hope first named it "strontianite" after the lead mines of Strontian in Southern Scotland, but later it was shortened to "strontium."[8] When Pleasant Valley's 4-H Club visited Davis, strontium would have been found in the cathode ray tubes of the television sets that illuminated their homes, preventing dangerous X-ray emissions. But as cathode ray tubes were superseded by liquid-crystal displays in recent decades, strontium's economic prominence diminished. Much of the mined element now rests within difficult-to-recycle TVs occupying landfills the world over.

But strontium, particularly strontium-90, also resides in the atmosphere. One of the earliest "planetary" tracer elements, according to Joseph Masco, strontium released during nuclear tests allowed earth scientists "to map a vast range of process[es], including weather systems, food chains, and environmental complexity in the early Cold War period."[9] The dispersal was a mixed blessing at best and disaster at worst, since strontium-90 penetrates aggressively into the body, replacing calcium in the bones.[10] The Atomic Energy Commission (AEC) already knew this early in the 1950s, starting a secretive project (codenamed "Sunshine") in 1953 to understand just how widespread the isotope had become. They kept the public in the dark about Sunshine, however.

In 1955, AEC Commissioner Willard F. Libby finally acknowledged strontium-90's presence in nuclear fallout, and its dangerous existence in the nation's milk was noted in government reports the following year. Medical journalist Walter Schneir refused to mince words in his exposé on the subject in *The Nation* in 1959: "People throughout the world will suffer death and illness from the nuclear tests conducted to date – and the effects of these tests will still be felt by mankind 10,000 years from now."[11] The "Sunshine unit," a lasting product of the period, is a metric that contrasts the amount of strontium per unit of calcium in a body. A body filled with Sunshine would, paradoxically, face extraordinary risks of cancer.

One year prior to Schneir's piece, activist-biologist Barry Commoner had co-founded the Greater St. Louis Citizens' Committee for Nuclear Information (CNI) in order to combat perceived government deception about the dangers of fallout from atmospheric nuclear testing.[12] Taking inspiration from Danish biochemist Herman M. Kalckar's proposal of an "An International Milk Teeth Radiation Census" and the United States Public Health Service's finding of elevated strontium-90 levels in St. Louis in 1958, the CNI decided to collect baby teeth and questionnaires concerning residency and dietary practices.[13] A brief blip in the *New York Times* the following March announced plans to "collect 50,000 baby teeth a year to measure the amount of strontium-90 absorbed by growing children."[14] The "Baby Tooth Survey," as the project became known, gradually surpassed these numbers, amassing two-hundred-thousand teeth by 1966. The CNI found that the amount of strontium in children's bodies, including those who saw the beagles at Davis, had "tripled and even quadrupled" from 1950 to 1954.[15] The sins of the body politic remained as traces in the human body – itself "the Ultimate Radioactive Storage Site," according to Kate Brown.[16]

Thus, when beagle dogs began eating strontium-laced dog food, they did so as representatives of both a possible future, laden with continuous nuclear tests and the promise of American dominance, and of an already-extant present, where milk from strontium-fed cows spread the substance widely in the food supply, a process that I have elsewhere called "species projection."[17] Researchers at Davis added radium-contaminated food to the beagles' diet in order to establish a proper human comparison, as radium exposure among dial painters represented the best understood radioactive reference point.[18] Yet when the Soviet Union, United Kingdom, and the United States signed the Partial Nuclear Test Ban Treaty in Moscow on 5 August 1963, the beagles were once again out of sync with geostrategic initiatives. Davis would search (and eventually find) alternative toxicity studies to enrol them in.

The Beagle Club studies that littered the United States between 1950 and 1990 are rarely discussed today.[19] They have no memorials, and their annual reports are scattered across university libraries and government archives, collecting the soft dust of inattention. One pile sits in my apartment, the gift of health physicist Otto Raabe, who planned to throw them away. In many ways, researchers anticipated this fate: a retrospective symposium on the projects in 1983 highlighted fears that the "data may not permit confident extrapolation to the human and may afford little in the way of basic understanding."[20] This data – including preserved tissue samples and meticulous logs catalogued on purpose-built computers – appeared irrelevant, waste from wasted resources which might do little more than "point to the fact that an effort was made to study the problem" and then disappear seemingly without a trace.

Yet the gallon drums holding carcasses still hot with strontium and radium, and the concrete blocks of shit snug underground in Washington, serve as alternative media of memory. As below, so above: Hanford's landscape, seen from buses that ferry visitors through this stygian "Manhattan Project National Historical Park," is studded with massive, enclosed reactor sarcophagi. When I visited in March 2017, I was overcome by the awesome emptiness of blank fields, hemmed in on all sides by hills, with little more than scrub brush and massive powerlines following the roads for miles – a characteristic intertwining of "wasteland" and "wilderness," according to Peter Galison.[21] Here, despite ongoing power production at the Columbia Generating Station, the enthusiasm that once drove researchers to deliver radioactive aerosols via masks to pigs and beagles feels distantly past. Except, that is, in the visual lexicon of local business: the Bombing Range Brewery's logo was a beer hop-"Fat Man" nuclear bomb, while the Midtown Dental Clinic was represented by an atomic tooth. The legacy of the Babytooth Survey is a strange one.

In *Containment*, Galison and Rob Moss suggest that nuclear storage invites planners to think on uncomfortable scales of 10,000 years that relativize the historical time on which much science and governance is predicated. Science fiction writers and futurists once joined together to predict how future peoples could be warned about the dangers, a tacit acknowledgment that "we" (a Western, male "we") were unlikely to be present. The situation was dramatically different, however, for Indigenous and Aboriginal peoples who had cared for the land over millions of years and expected to do so for millions more. Though the beagles invited less potent imagination, the inhuman durability of their waste exemplifies one of the more pressing challenges of life within the period colloquially called the "Anthropocene": the problem of perpetual containment and the impossibility of elimination; the survival of the trace.[22] And as Gabrielle Hecht reminds us, "the violence associated with the Anthropocenic apotheosis of waste ... has particular, differential manifestations."[23] Such waste, whether as dog drums or waste drums, inactive reactors or perpetual storage sites, remains an insistent, Chthonic memory liable to outlast many of its caretakers, transfiguring the land with toxicity.

Where strontium-90 first demarcated the nuclear age atmospherically, it re-emerged more recently in the "living laboratory" produced by the Fukushima meltdown.[24] Between the seeming infinities of radioactive half-lives and the evanescence of experimental life lies another medium of persistence: the archive or database. "The Beagle Dog Tissue Archive," hosted by Northwestern University's Woloschak Lab, holds information from the Beagle Club studies for digital retrieval – potentially invaluable for contemporary researchers unlikely to receive permission for another study of such size and scale given restrictive laboratory animal laws. In a recent paper, members of the lab grouped atomic bomb survivors and uranium miners together with exposed mice and beagles as so many surviving "radiation data archives."[25] Irradiation makes life forms into forms of archive, demonstrating Achille Mbembe's suggestion that the archive is "not a piece of data, but a status."[26]

The beagle data, including uncatalogued HTML lists of scanned documents I have monitored for years, is a lingering monument to the disappointments of twenty-first-century scientific internationalism. The American National Radiobiology Archives (NRA) and European Radiobiological Archive (ERA), intended as caretakers for data that was "not only expensive but also of doubtful value because of the risk of deterioration of slides and tissue blocks," have given up or cut back in the face of funding shortfalls and alternative priorities.[27] As a team of researchers for the ERA put it, preservation of the data remains critical if the thousands of tissues stored at Northwestern and in Germany are to become more than miniature tombs in an "information graveyard" accompanying the radioactive pet cemetery at Hanford and the newspaper mausoleums online.[28] A reversal is unlikely under America's current administrations, so the Beagle Dog Tissue Archive is apt to remain an optimistic relic.

These efforts at Davis and elsewhere – what Geoffrey Bowker calls "memory practices" – could not guarantee the utility of their data, nor ensure that the knowledge would survive.[29] As journalist Linda Nordling noted in 2010, "Digital information collected between 1950 and 1980 is also threatened, because it is stored on outdated media often subject to deterioration, such as magnetic tape and floppy disks, making it increasingly difficult to access and retrieve."[30] But as the beagles demonstrate, austerity policies are just as threatening as retrograde media. Scientific memory, like much of human life in capitalist ruins, appears more fragile than anticipated. As the Pleasant Grove women enter their 70s, results join threatened species as subjects of the twenty-first century's pervasive "endangerment sensibility."[31] Yet the pace of scientific production hardly slows, and with the publication of ever more results, studies become waste. In turn, waste becomes studied, contained, and cleansed by risk-adverse administrative bodies. As Bauman writes, "Recycling of waste is itself a waste-producing" – or trace-producing – "process."[32]

Of equal concern are the practical consequences of a revelation that historians and anthropologists of science have trumpeted for decades: the "hard truths" of research are no longer so effortlessly confirmed when the results that might back them up cannot be located, let alone read. Databases, imagined to permanently stabilize scientific memory, that obscure object of desire, during the long march of perpetual progress, are themselves ephemeral, liable to wash away in political and environmental tides – liable to become waste.

Although the "contaminated ghost town of empty dog kennels" is gone, the beagle tissues remain, a persistence of more-than-human forces and animal traces as human achievements lose their lustre.[33] And even then, the past continues to haunt the present: after Fukushima, scientists once again began collecting teeth to mark the spread of contamination (this time from cows), and a beagle mix was rescued from the Kido Japanese Railways Station at the southern edge of the exclusion zone.[34] While other residues of "big" radiobiology are scoured away, and the Anthropocene supersedes Nuclear War as the unpredictable and apocalyptic sum of all fears, the return of the elemental reveals the complicated alter-lives of traces.[35] There is, at least, one certainty: the shit will stick around.

NOTES

1   "Beagle Project at Davis Seen By 4-H Members," *Appeal-Democrat*, 11 April 1961.
2   Hans-Jörg Rheinberger, *Toward a History of Epistemic Things: Synthesizing Proteins in the Test Tube*, Writing Science (Stanford: Stanford University Press, 1997), 4.
3   Etienne Benson, "Animal Writes: Historiography, Disciplinarity, and the Animal Trace," in *Making Animal Meaning*, eds. Linda Kalof and Georgina M. Montgomery (East Lansing, MI: Michigan State University Press, 2012), 6.
4   Dale Kasler, "UC Davis Irradiated Beagles during the Cold War: Now It's Agreed to Clean the Laboratory Landfill," *The Sacramento Bee*, October 5, 2018, sec. Education, https://www.sacbee.com/news/local/education/article219515130.html.
5   Eva Giraud and Gregory Hollin, "Care, Laboratory Beagles and Affective Utopia," *Theory, Culture & Society* 33, no. 4 (July 2016): 27–49, https://doi.org/10.1177/0263276415619685; Brad Bolman, "How Experiments Age: Gerontology, Beagles, and Species Projection at Davis," *Social Studies of Science* 48, no. 2 (2018): 232–58.
6   Associated Press, "Hanford Dogged by New Problem of Waste Disposal," *The Register-Guard*, October 16, 1990.
7   Zygmunt Bauman, "The Sweet Scent of Decomposition," in *Forget Baudrillard?*, eds. Chris Rojek and Bryan S. Turner (New York: Routledge, 1993), 39; Dominique Laporte, *History of Shit*, trans. Nadia Benabid and Rodolphe el-Khoury (Cambridge, MA: The MIT Press, 2000).
8   François-Xavier Coudert, "Strontium's Scarlet Sparkles," *Nature Chemistry* 7, no. 11 (November 2015): 940, https://doi.org/10.1038/nchem.2376.
9   Joseph Masco, "The Age of Fallout," *History of the Present* 5, no. 2 (Fall 2015): 147.
10   Laura A. Bruno, "The Bequest of the Nuclear Battlefield: Science, Nature, and the Atom during the First Decade of the Cold War," *Historical Studies in the Physical and Biological Sciences* 33, no. 2 (2003): 241, https://doi.org/10.1525/hsps.2003.33.2.237.
11   Walter Schneir, "On the H-Bomb Front: 1: Strontium-90 in U. S. Children," *The Nation*, April 25, 1959, 355.
12   Michael Egan, *Barry Commoner and the Science of Survival: The Remaking of American Environmentalism* (Cambridge, MA: The MIT Press, 2007), 60.
13   Egan, *Barry Commoner*, 67.
14   UPI, "Teeth to Measure Fall-Out," *The New York Times*, March 19, 1959.
15   Egan, *Barry Commoner*, 71.
16   Kate Brown, "The Last Sink: The Human Body as the Ultimate Radioactive Storage Site," *RCC Perspectives: Transformations in Environment and Society* 1 (2016): 41–7, https://doi.org/10.5282/rcc/7392.

17 Bolman, "How Experiments Age," 10.
18 Luis A. Campos shows how the dial painters were seen as part of radium's dissociation with life-giving powers. See Luis A. Campos, *Radium and the Secret of Life* (Chicago, IL: The University of Chicago Press, 2015), 242–3.
19 Beyond UC-Davis, beagles were utilized in radiation research at the University of Utah, the Hanford Works (later Pacific Northwest National Laboratory), the Argonne National Laboratory, the Lovelace Institute, the University of Rochester, Colorado State University, and others.
20 Roy C. Thompson and Judy A. Mahaffey, "Preface," in *Life-Span Radiation Effects Studies in Animals: What Can They Tell Us? Proceedings of the Twenty-Second Hanford Life Sciences Symposium Held at Richland, Washington, September 27–29, 1983* (Springfield, Virginia: Office of Scientific and Technical Information, 1986), ix.
21 Cited in Başak Saraç-Lesavre, "Desire for the 'Worst': Extending Nuclear Attachments in Southeastern New Mexico," *Environment and Planning D: Society and Space*, November 26, 2019, 10, https://doi.org/10/ggjvcp.
22 Emily Simmonds, "Nuclear State, Nuclear Waste," interview by Max Liboiron, November 5, 2018, https://discardstudies.com/2018/11/05/nuclear-state-nuclear-waste/.
23 Gabrielle Hecht, "Interscalar Vehicles for an African Anthropocene: On Waste, Temporality, and Violence," *Cultural Anthropology* 33, no. 1 (2018): 112.
24 For post-meltdown spaces as zones of experimentation, see Adriana Petryna, *Life Exposed: Biological Citizens after Chernobyl*, with a new introduction by the author (Princeton, NJ: Princeton University Press, 2013); Kate Brown, *Manual for Survival: A Chernobyl Guide to the Future* (New York, NY: W.W. Norton & Company, 2019). For Fukushima specifically, see Georg Steinhauser, Viktoria Schauer, and Katsumi Shozugawa, "Concentration of Strontium-90 at Selected Hot Spots in Japan," *PloS One* 8, no. 3 (March 2013).
25 Alia Zander, Tatjana Paunesku, and Gayle Woloschak, "Radiation Databases and Archives – Examples and Comparisons," *International Journal of Radiation Biology* 95, no. 10 (October 3, 2019): 1379, https://doi.org/10/ggjvfr.
26 Achille Mbembe, "The Power of the Archive and Its Limits," in *Refiguring the Archive*, ed. Carolyn Hamilton et al. (Dordrecht, NE: Kluwer Academic Publishers, 2002), 20.
27 G.B. Gerber et al., "International Radiobiology Archives of Long-Term Animal Studies: Structure, Possible Uses and Potential Extension," *Radiation and Environmental Biophysics* 38 (1999): 76.
28 Paul N. Schofield, Soile Tapio, and Bernd Grosche, "Archiving Lessons from Radiobiology," *Nature* 468, no. 7324 (December 2, 2010): 634, https://doi.org/10.1038/468634a.
29 Geoffrey C. Bowker, *Memory Practices in the Sciences* (Cambridge, MA: The MIT Press, 2008).

30 Linda Nordling, "Researchers Launch Hunt for Endangered Data," *Nature News* 468, no. 7320 (November 2, 2010): 17, https://doi.org/10.1038/468017a.

31 Fernando Vidal and Nélia Dias, "Introduction: The Endangerment Sensibility," in *Endangerment, Biodiversity and Culture*, eds. Fernando Vidal and Nélia Dias (New York, NY: Routledge, 2016), 2.

32 Bauman, "The Sweet Scent of Decomposition," 39.

33 Richard C. Paddock, "Fallout From Beagle Experiments," *Los Angeles Times*, February 8, 1994.

34 Kazuma Koarai et al., "^{90}Sr in Teeth of Cattle Abandoned in Evacuation Zone: Record of Pollution from the Fukushima-Daiichi Nuclear Power Plant Accident," *Scientific Reports* 6, no. 1 (April 5, 2016): 1–9, https://doi.org/10/ggjvgx; Mayumi Itoh, *Animals and the Fukushima Nuclear Disaster* (New York, NY: Palgrave Macmillan, 2018), 81.

35 Michelle Murphy, "Alterlife and Decolonial Chemical Relations," *Cultural Anthropology* 32, no. 4 (2017): 497.

14 TECTONICS

zeynep oguz

On 17 August 1999, a violent earthquake shook Turkey's north-western regions, killing 17,000 people. Dubbed by Turkish news media and politicians as a national tragedy, the 1999 Izmit earthquake brought two things to public attention. First, it exposed the massive scale of corruption that had taken root in various levels of government and the construction sector in Turkey. For many, it was clear from the earthquake's impacts that this was no "natural" disaster but a predominantly political one caused by faulty buildings, corrupt contractors, and the state institutions that ignored them. Second, it turned ordinary citizens into tectonic and seismic quasi-experts. As geologists became regular TV show guests, audiences became well-versed in magnitude scales, aftershocks, and earthquake preparedness. Plate tectonics and faulting in Turkey were at the centre of the media spectacle. For the first time, publics learned that Turkey sits on the boundary of three tectonic plates – Eurasian, Arabian, and Anatolian – and two major active fault lines – the North Anatolian Fault and the East Anatolian Fault – and that the 1999 earthquake was caused by seismic activity in the former. It was evident that an earthquake of such massive impact could not just happen anywhere. Instead, the earthquake exposed the unique relationship between political and geological forces in Turkey, where neither had final determining power.

In this chapter, I explore the specific nature of this relationship, tracing how geological and geophysical forces also *move* political and social formations. Through this tectonic lens, Turkey – a former multi-ethnic and multi-linguistic empire – is revealed to be a convoluted site of imperial collapse and nation-building, colonial and territorial politics, and anti-colonial insurgency that spans from the end of the First World War to the present. Inspired by calls to decolonize the Anthropocene on the one hand, and provocations to rethink the counter-hegemonic potential of the geological on the other, I explore the decolonial work that an attention to plate tectonics might do in Turkey, where Kurdish people have been historically subjected to violence and oppression.[1] I argue that attention to such elemental forces as plate tectonics can play a role in modes of what political philosopher Walter Mignolo calls "epistemic disobedience" that unsettle colonial and violent forms of power in the Anthropocene.[2]

The story of plate tectonics began around 200 million years ago, when the supercontinent Pangaea broke up into two continents, starting a mutation that eventually formed Earth's present-day continents.[3] This continental emergence was driven by the motion of tectonic plates, fractured lithospheric formations of Earth's outermost crust which glide and move alongside each other, propelled by the thermal currents in the near-molten mantle. Earth scientists contend that there are three different types of plate boundary: divergent plate boundaries, when plates move away from each other; transform boundaries, where plates just slide past one another; and, finally, convergent boundaries, where plates collide. Plates are either dragged back down oceanic trenches into the mantle beneath, or crumpled, deformed and pushed up into new mountain ranges, slowly renovating and renewing the lithosphere.

Plate tectonics, as the "unifying theory of modern geology," cast the Earth as a dynamic and multi-layered composition, describing the mobile character of the planet's surface features and explaining the formation of mountains, trenches, volcanism, and seismicity.[4] Earth scientists see mantle convection and ocean subduction as a significant factor that has affected planetary habitability and the evolution of life in the oceans through its ability to exchange and circulate materials such as zinc, copper, selenium, and cobalt that have been essential for the evolution of early life forms.[5] Plate tectonics is thus a theory of relationality, a framework that explains the interscalar linkages between physical forces of heat and pressure across different scales of space and time. But as the 1999 earthquake illustrates, the relations that tectonic plate movements bring about are not merely geophysical or geological but also inherently political. In the following sections, I use the concept of "geopower" as theorized by the feminist philosopher Elizabeth Grosz to explain the specific workings of this entanglement.

Geopower, for Grosz, designates the relations between Earth and its political, social, and territorial formations.[6] The geological and geophysical forces of the Earth are, in this view, subject to the workings of political power, just as political events and formations are also conditioned by the geological. Thus, one of the critical insights of geopower is that the relationship between political and geological forces does not flow in a single direction but is *multidirectional*; while political actors strive to territorialize Earth, the workings of power are also conditioned by earthly forces. Merely recognizing the multidirectional relationship between the Earth and power, however, fails to explain the social and political worlds that are conjured up through this relationship. Understanding the workings of this multidirectional structure may further open up new possibilities that subvert hegemonic and violent forms of power. Turkey's specific tectonic setting allows this geo-political collision to become a productive analytic, allowing us to trace entanglements through a variety of geological formations, themselves brought about through what geologists call "compressional stress," such as crustal deformations, fault lines and mountains.

This regional tectonic story began 90 million years ago when the African and Eurasian plates began to collide. Around 25 million years ago, their collision closed the oceans between them. The Anatolian Plate, which the majority of present-day Turkey's landmass rides upon, is located in between the convergent boundary of these two giant colliding plates. Being compressed and pushed west by the Arabian Plate and pulled apart on its western margin, where it meets the Aegean Sea, the Anatolian Plate has been slowly rotating counterclockwise since this collision event.[7] The intracontinental collision has sutured the Anatolian plate along the Bitlis-Zagros Suture Zone, which has formed as a part of the larger Alpide-Himalayan orogenic belt, while thickening the crust and uplifting eastern Anatolia. Today, Turkey's current topography is characterized by two major fault lines and widespread mountain ranges. I take these geological formations not only as objects of anthropological attention, but also as analytic vehicles that reveal the composition of political and social worlds.

Two concepts help capture the relationship between tectonic plates and political power. The first is "interscalar vehicles" which, for the historian of science Gabrielle Hecht, are "objects and modes of analysis that permit scholars and their subjects to move simultaneously through deep time and human time, through geological space and political space."[8] Hecht writes that "to grapple with the complex interscalar connections posited by the Anthropocene, we need new narratives and analytic modes,"[9] using elements such as uranium to examine relationships across scales, including the planetary temporalities that the Anthropocene brings into view. The second concept is Sandro Mezzadra and Brett Neilson's "border as method." Taking political borders not only as a "research 'object' but also as an 'epistemic angle'" in their analysis of living labour across diverse geographical scales, they bring into view unexpected reverberations between concepts, histories, and materialities that otherwise seem distant from each other.[10] This is an essential task in the Anthropocene, which requires methodologies and objects of study that bring together "stories and scales usually kept apart."[11] As Hecht shows, such a move can piece together seemingly disparate histories of Cold War technoscience, mining and coloniality in Africa, industrial waste, and deep time.

In the case of Turkey, when one takes Earth not just as an ethnographic object, but also as an analytical medium to capture the conditions of the Anthropocene, the dissonances and resonances of state and region-making, extractivism, post-imperial imaginaries, and territorial (and counter-territorial) politics collide alongside the elemental forces of tectonic collision. Tectonics are elemental because they are constitutive of the stage where the political and social activities of humans take place on Earth. At the same time, they become the literal strata through which states attempt to naturalize their claims over land. Yet the shifting forces of tectonics also become the grounds for the pursuit of a politics of the otherwise – one that disrupts and subverts the violent grip of the nation-state, capitalist extractivism, and their war machines over land, peoples, and more-than-human life. In this way, taking tectonics as elemental forces affirms this volume's claim to the elemental as "both methodological and political."[12] If geological and geophysical forces underpin social and political formations across multiple scales, taking *tectonic plates as method* may work to reveal the "forms of geopower or arrangements of Earth forces make territorialization possible in the first instance."[13]

Taking plate tectonics as both an ethnographic object and an analytical device, the next sections explore the entanglement of faults and folds, mountains and the underground, with oil extraction, territorial politics, and warfare; with imperial and national histories, deep time, and imaginaries of resource abundance and territorial expansion.

As noted above, the Eurasian-Arabian collision has generated two major sutures: the North Anatolian Fault Zone (NAF), which is a 500-kilometre broad arc-shaped fault system that extends from eastern Turkey to Greece in the west, formed in the Early Pliocene (5.3 to 3.6 million years ago), and the East Anatolian Fault Zone (EAF), a 550-kilometer northeast-trending fault zone, formed in the Late Pliocene. The North Anatolia Fault separates the Eurasian Plate from the Anatolian Plate in northern Turkey, and the Eastern Anatolian Fault separates the Arabian Plate from the Anatolian Plate. Living atop these relations, Turkey is thus one of the most seismically active regions in the world, having a long history of massive and devastating earthquakes. Here, however, I do not trace seismicity and earthquakes – which are relatively more direct and eventful manifestations of plate tectonics[14] – but instead travel with geo-political formations that are more temporally and spatially dispersed, such as the geopolitical life of oil and mountains in Turkey's Kurdish-populated southeast.

The faulting and folding of the crust in this collision zone have directly impacted the predicament of oil and its exploration in Turkey. While Turkey's southeast constitutes the northernmost edge of the oil-rich Arabian Plate, the Anatolian Plate is geologically a separate entity and, therefore, despite its intimate proximity, the later plate does not have the same sedimentary and source rocks that trap large oil deposits in the Arabian Plate. Earth's crust has been significantly deformed by the Eurasian-Arabian collision here, as the continental crust buckled, and rocks piled up, lifting this crust and forming mountains. Geologists assume that oil has either completely escaped through the fractures created in rocks or settled in tiny traps due to this process, making its petrochemical potential both extremely difficult to find, and, if discovered, often commercially unviable.

Nonetheless, in the 1970s, Turkish geologists discovered small- to medium-sized oil deposits in the regions surrounding the south-eastern district of Diyarbakir. More than four decades later, in the spring of 2017, I was in a village near Diyarbakir following a Turkish exploration geologist, himself employed by the state-owned Turkish Petroleum Company to discover further deposits. We were accompanied by the village chief, a Kurdish man in his forties, who asked a very legitimate question of the geologist: "Why do we have so little oil whereas right below us, in Iraq, oil is literally blowing out from under the ground? It doesn't make much sense to me." Quite content with having the chance to explain his favourite subject, the geologist replied, "Well, it does, because the southeast [of Turkey] is geologically very different from Iraq." He went on:

> Imagine a room covered with a rug from one end to the other. What happens when you open the door? The part of the rug that's close to the door buckles and folds, right? So, this buckled and folded part of the rug is like south-eastern Turkey, and the far, straight corner is like the Arabian Peninsula. It's much easier to explore, discover, and produce oil in there. This is also why, we say, in contrast to our neighbours in the Middle East, we don't have much oil.

Indeed, Turkey's domestic oil reserves supplied only 7 per cent of the country's oil consumption in 2016.[15] Currently, oil production in Turkey is around 50–60 thousand barrels per day, a meagre amount compared to Iran or Iraq, right near Turkey's southern and eastern borders, which each produce 3 to 4 million barrels per day. Rumours and anticipation about possible oil abundance have been a key part of Turkey's sociopolitical landscape since the beginning of state-led oil exploration efforts in south-eastern and eastern parts of the newly established nation-state from the ruins of the Ottoman Empire and in the aftermath of the First World War. Oil manifests, socially, in an abundance of conspiracy theories about its elusive presence. According to a widespread urban myth in Turkey, for example, the 1923 Treaty of Lausanne – the document internationally recognizing Turkey in the aftermath of the 1919–23 national war of independence – will soon expire upon its hundredth anniversary, rendering Turkey's current borders obsolete. Proponents of this theory claim that the treaty includes secret clauses that have prevented Turkey from having full sovereignty over its resources and thereby extracting its supposedly abundant oil reserves.

Until 2019, the Turkish Petroleum Company website's "Frequently Asked Questions" section was largely devoted to refuting such conspiracy theories. "Why does Turkey have no oil when its neighbours have the largest reserves in the world?" it asked. "Is Turkey floating on a sea of petroleum?" Answering such questions, the website refuted conspiracy theories and explained why, despite neighbouring rich oilfields, Turkey has relatively poor oil resources, with references to plate tectonics, the Arabian-Eurasian collision, and the formation of the Alpine-Himalayan Belt. Yet these magical theories stick in public life, as they often work to ignite nostalgia about the extended borders of the former Ottoman Empire, fuelling the desires for territorial expansion that characterize nationalist politics in contemporary Turkey. President Erdogan, for example, often alludes to the former or "lost" territories of the Ottoman Empire, and the potential malleability of the Treaty of Lausanne, as legitimizing factors in Turkey's military operations in Iraq and Syria, and its irredentist aspirations in the Middle East. The heavy faulting that has occurred due to the specific workings of plate tectonics in Turkey's eastern regions, therefore, conjures up fantasies of geo-political power that the current authoritarian and populist Justice and Development Party (AKP) regime capitalizes upon. The geological, here, is constitutive of political power in contemporary Turkey, but as the next section demonstrates, the territorial and military capacities of state power and sovereignty have also been significantly constrained by the geological and geophysical manifestations of plate tectonics.

Taking tectonic plates as method requires an attention to the plate boundaries themselves, tracing the frictions between plate boundaries and political boundaries, borders and regions. Turkey might not be rich in oil, but almost all of its oil is drilled in south-eastern regions where the Arabian

Plate slides under the Anatolian plate, elevating the landmass, faulting and folding the strata. For the Turkish state, this geography encompasses the administrative region known as South-Eastern Anatolia, a designation given during the first national Geography Congress held in 1941. The congress was one of the many early attempts by the new Turkish Republic to facilitate the territorialization of Anatolia – one of the few landmasses left following the demise of the Ottoman Empire – as a national space composed of a culturally homogenous and Turkified population.

But South-Eastern Anatolia is not the only known name for this region. For millions of Kurds in Turkey and beyond, Anatolia's south-east is known as Northern Kurdistan – an unacknowledged "internal colony" since the nineteenth century and a contemporary zone of late-modern colonial occupation.[16] Prior to the formation of the Republic of Turkey in 1923, "Kurdistan" (or "Land of the Kurds") denoted the geographical area of Kurdish settlement that roughly included the mountain systems of the Zagros and the eastern extension of the Taurus. Ottoman centralization policies in the nineteenth century disrupted the relative autonomy of Kurdistan, suppressed the power of local Kurdish leaders, and dismantled the structure of Kurdish emirates. Multiple rebellions followed but were violently suppressed by the Ottoman armies and their successors.[17] After the First World War, and following the principles of self-determination, the Allied powers promised the Kurdish people their homeland in the 1920 agreement known as the Treaty of Sèvres. But the subsequent Treaty of Lausanne, following the occupation of Turkey and a national war of independence, dismissed such prior promises of autonomy and instead divided the Kurdish people among Turkey, Iraq, Syria, and Iran.

With the collapse of Kurdish independence and the establishment of the Turkish Republic in 1923, the Kurdish people in Turkey faced violent acts of cultural assimilation and political repression. The young nation-state officially denied the existence of Kurdish identity and perceived Kurdish language and culture as an existential threat to the territorial unity of the new nation. Kurds in Turkey's south-eastern regions were forcefully deported from their lands and assimilated into Turkish identity, while the region – Northern Kurdistan– was ruled with a heavy military presence and was systematically underdeveloped. But, despite these oppressive efforts, the Turkish state failed to incorporate the Kurdish people and Kurdistan into its national space.

Since 1984, and the rise of an armed insurgency led by the guerilla organization Kurdistan Workers' Party (PKK), this region has been a war zone. The PKK adopted armed struggle against the Turkish military as a revolutionary strategy for cultural and political rights and self-determination for Kurds and, in the early 1990s, this insurgency evolved into a mass movement with several million supporters and sympathizers drawn from all parts of Kurdistan. The Turkish state retaliated by declaring a state of emergency in 11 provinces, conducting massive military operations in mountain hideouts, villages and cities, and adopting tactics of authoritarian violence such as paramilitary attacks, forced disappearance and torture.[18] Until 2002, Northern Kurdistan officially remained under martial law and state-of-emergency rule. The mountains of the nation's southeast hosted guerrillas for decades, and while they have always been important in Kurdish folklore, they attained a particular significance in this period of violence and warfare, representing sanctuary, hope, and connection for the divided Kurdish people.

The final manifestation of the Eurasian-Arabian tectonic collision I trace here is the Zagros mountains, also known as the Zagros fold and thrust belt (or "Zagros FTB"). The Zagros is a wide mountain range that begins in north-western Iran and follows its western border while covering much of south-eastern Turkey and north-eastern Iraq, and thus most of Kurdistan. The mountain ranges in Northern Kurdistan, where the PKK fought against Turkish soldiers, are extensions of the Zagros range in Turkey's south-east. Since the late 1980s, these mountains have been geological mediums through which insurgency and counterinsurgency are waged.

Quoting the famous Kurdish proverb that "the Kurds have no friends but the mountains," political scientist Michael Gunter notes in his reporting on "the Kurdish question" that the mountains are the "most prominent geographic characteristic of landlocked Kurdistan."[19] For Gunter, the rugged and mountainous terrain is one of the primary factors obstructing the Kurdish peoples' political unity. He argues the terrain has kept Kurdish tribes apart from each other and delayed them from developing a collective or national identity. Yet, for Gunter, the same mountain ranges in the region have also protected the Kurds from being fully conquered or assimilated by Turkish, Persian, and Arab states. Similarly, sociologist Akın Ünver writes that the Zagros mountains have "both rendered the Kurds extremely resilient to systemic changes to larger states in their environment, and also provided hindrance to the materialization of a unified Kurdish political will."[20]

As the war between the Turkish Armed Forces and the PKK continued, the Turkish state and the national media eventually realized that the mountains and the underlying tectonic plates held political and territorial significance. In the 2000s, the Eurasian-Arabian collision became the subject of ongoing public discussion and news, appearing, for example, in the politics section of a national newspaper in 2007. The popular pro-state Turkish daily *Milliyet* (*Nationhood*) published an article entitled "Geologic Collision in the Southeast!" with the subheading:

"Nature Provides Convenience to Terrorists." Drawing on the statements of a Turkish geologist and the Turkish army's recent counterinsurgency reports in the region, the article formed a direct link between the persistent tectonic collision and political insurgency:

> Southeast Turkey has a mountainous structure, due to the collision of the Arabian Peninsula with the Anatolian Plate. According to scientists, the geological structure that allows terrorists to take shelter and go unnoticed easily is the result of a geological process during which the Arabian Peninsula's collision with Anatolia left southeast Turkey compressed between them. The land was, thus, elevated, creating a steep, rocky and mountainous geography.[21]

By linking these two phenomena – tectonics and armed conflict – the article aligned inhuman scales of time and space with human ones. It tied the movement of Earth's lithospheric plates to the ongoing war between the Turkish army and the PKK. Reading this article, critically and symptomatically, while tracing the formation of Kurdistan's mountains back to the Eurasian-Arabian tectonic collision, problematizes geological formations as naturalized objects and mediums of state power. Nation-making projects tend to catalyze physical characteristics of the land as natural boundaries or the "geobody" of the nation-state.[22] Following the establishment of the Republic of Turkey in 1923, nationalist geographers and historians strived to depict Anatolia as a natural-geological boundary of the new Turkish Republic. Yet northern Kurdistan's geological difference from the Anatolian Plate renders this narrative problematic, as it is a part of the Arabian Plate. Such a task also reveals that geological formations and processes may work as unsettling forces that become mediums of insurgency. Anti-hegemonic actors may also utilize geological formations to unsettle the territorial and colonial politics of the nation-state.

The accounts outlined above point out the ways in which the geological manifestations of plate tectonics have been capitalized by the politically powerful, enabling them to provoke political imaginaries of oil wealth and territorial sovereignty. Yet as I have demonstrated, they have also rendered such hegemonic imaginaries problematic, at times limiting them with their very materiality, and at other times, becoming vectors through which decolonial actors embark upon counter-territorial practices. As the next section shows, plate tectonics might further work to recompose worlds otherwise.

In a critical decolonial intervention into geopolitics, geographer Angela Last asks, "What role might geophysical forces play in challenging hegemonic geopolitical worldviews?"[23] Tectonic plate movements and the geological formations that they have manifested in Turkey reveal a constellation of resource extraction, colonial politics, and post-imperial territorial power today. Yet an attention to the geologic might further work to align with Last's decolonial and anti-hegemonic agenda and disarray this violent constellation. To return to Mezzadra and Neilson, taking border as method is more than methodological, in that it foregrounds questions of politics that pertain to "understanding the social worlds produced at the border and intervening in those worlds," opening up a "space in which a different imagination and production of the world become possible."[24] In anthropology, questions of methodology have never been understood as a mere matter of selecting tools of empirical research. Instead, methodological questions have been entangled with interventions into the social composition of things and worlds in their broadest sense; as interventions through praxis, "ontological experiments" that might prove useful for cultivating other modes of existence.[25]

In Turkey, the geological has played a critical role in both enabling and disrupting the extractive, colonial, and territorial projects of the Turkish state, revealing how naturalized political regions and territories are enabled, but also limited by geological forces in the first place. Turkey's political geography is closely tied to its geological disunity, sitting between two huge tectonic plates that have been inexorably grinding against one another for millions of years. The convergent plate boundary where northern Kurdistan sits upon is also a collision zone of the elemental forces of *geos* and power, where diverse histories of oil, coloniality, warfare, forms of political imagination, resistance, refusal, and aspiration are lived. Efforts to territorialize the substrata sometimes fail, as the materiality of the terrain is always dynamic and fluent; violent colonial politics are unsettled, and finally, alternative political imaginaries are forged through the elemental medium of the geological. As the Turkish case demonstrates, geological plate boundaries rarely comply with human political territories and areas, negating claims of geological and national unity. Petroleum escapes through sutures, rendering ambitions of oil extraction obsolete. The frictions that take place in this geopolitical collision zone bring existing territorial orders, colonial regimes, and extractive politics into question.

The counter-territorial openings revealed by the elemental force of tectonics are closely linked to the broader politics of the decolonial project of the Kurdish freedom movement. In Syria, in collaboration with local communities – Arabs, Armenians, Assyrians, and Turkmens – the Kurds who have led the Rojava Revolution have been experimenting with democratic confederalism and autonomy. Inspired by the social ecology and libertarian-anarchist socialism of Murray Bookchin, the Kurdish freedom movement places concerns of ecological justice at the centre of political and social struggles. In Turkey's Kurdish cities, up until the Turkish state's violent crackdown in 2015, pro-Kurdish municipalities implemented cultural decolonization tactics to reappropriate urban space and reclaim that sense of belonging through self-governance, self-production, self-creation and self-defence.[26] Finally, cartographic representations of Greater Kurdistan – such as a stamp that depicted the map of Kurdistan and gifted to Pope Francis by Iraq's Kurdistan Regional Government (KRG)[27] – have emerged as powerful counter-artefacts that promote an alternative geological and geographical imaginary to the existing territorial orders.[28] A decolonial praxis, in these politics that have characterized the Kurdish freedom movement in the past decade, designates a set of discourses and material practices that aim to create spaces freed from the colonial violence and ecocide of the nation-state and capitalist extractivism. It refers to both countering the oppression of the Kurdish people and Kurdistan, an "international colony" distributed across four nation-states and composing alternative spaces and territorial imaginaries.[29]

The decolonial aesthetic practices of Kurdish artistic production are also central to the cultivation of counter-territorial imaginaries in Kurdistan today. The Kurdish-Iraqi artist Walid Siti's *A Poem to the Mountain at the Edge of the World*[30] (see figure 14.1) captures the potentials of the geologic in subverting existing territorial regimes and imagining alternatives. His 70-inches tall and 120-inches wide mixed-media sculpture is a recomposition of the iconic Zagros Mountains of Kurdistan. Rocks, mountains, and other earthly matter populate Ziti's work, as having provided "sanctuary from submission, massacre, division and destruction" to Kurdish people for centuries.[31] The mountains in *A Poem* are geological formations in the midst of the politically divided landscape among four states. *A Poem* reimagines the Zagros Mountains as a single formation, in the absence of political borders and territories that make up Turkey, Iran, Iraq, and Syria today. In the absence of political borders and denominations, the negative aesthetics of *A Poem* recomposes the Zagros as a unity and transforms Kurdistan into an imaginary *and* material place that spans across different times and places. Rather than performing a "counter-cartography," however, *A Poem* constructs what I conceptualize as a "counter-geology," where subversion does not stem from producing alternative symbolic forms or epistemic analyses, but from working with the elemental forces of the Earth.

Territorial borders may claim a naturalized link between Earth and state sovereignty, but the very materiality of geology as a vector might work to denaturalize these violent and oppressive arrangements. Combining an elemental perspective with what Mignolo calls "critical border thinking"[32] can utilize the geophysical as a decolonial strategy and fold

Figure 14.1. Walid Siti, *A Poem to the Mountain at the Edge of the World*, 2019

it together with Last's provocative question about the decolonial and anti-hegemonic possibilities of the geophysical. As the constitutive powers and subversive potentials of tectonics reveal, contemporary colonial relations and their territorial orders might be unsettled through sustained attention to the political possibilities of the elemental.

NOTES

1 Heather Davis and Zoe Todd, "On the Importance of a Date, or Decolonizing the Anthropocene," *ACME: An International E-Journal for Critical Geographies* 16, no. 4 (2017); Angela Last, "Fruit of the Cyclone: Undoing Geopolitics through Geopoetics," *Geoforum* 64 (2015).

2 Walter D. Mignolo, "Geopolitics of Sensing and Knowing: On (De)Coloniality, Border Thinking and Epistemic Disobedience," *Postcolonial Studies* 14, no. 3 (2011).

3 Henry R. Frankel, *The Continental Drift Controversy: Volume 1, Wegener and the Early Debate* (Cambridge, UK: Cambridge University Press, 2017).

4 Naomi Oreskes, *The Rejection of Continental Drift: Theory and Method in American Earth Science* (Oxford, UK: Oxford University Press, 1999), 9; Frankel, *Continental Drift Controversy*, xv.

5 Rebecca Boyle, "Why Earth's Cracked Crust May Be Essential for Life," *Quanta Magazine.* (June 7, 2018), https://www.quantamagazine.org /plate-tectonics-may-be-essential-for-life-20180607/.

6 Elizabeth Grosz, Kathryn Yusoff, and Nigel Clark, "An Interview with Elizabeth Grosz: Geopower, Inhumanism and the Biopolitical," *Theory, Culture & Society* 34, no. 2–3 (2017); Kathryn Yusoff, "Geopower," in *Handbook on the Geographies of Power,* eds. Mat Coleman and John Agnew (Edward Elgar Publishing, 2018).

7 Jonathan C. Aitchison, Jason R. Ali, and Aileen M. Davis, "When and Where Did India and Asia Collide?," *Journal of Geophysical Research: Solid Earth* 112, no. B5 (2007).

8 Gabrielle Hecht, "Interscalar Vehicles for an African Anthropocene: On Waste, Temporality, and Violence," *Cultural Anthropology* 33, no. 1 (2018): 135.

9 Ibid., 112.

10 Sandro Mezzadra and Brett Neilson, *Border as Method, or, the Multiplication of Labor* (Durham, NC: Duke University Press, 2013), viii.

11 Alison Kenner, Aftab Mirzaei, and Christy Spackman, "Breathing in the Anthropocene: Thinking through Scale with Containment Technologies," *Cultural Studies Review* 25, no. 2 (2019): 153.

12 Timothy Neale, Courtney Addison, and Thao Phan, introduction to *The Anthropogenic Table of Elements,* this volume.

13 Yusoff, "Geopower," 210.

14 Elizabeth Angell, "Assembling Disaster: Earthquakes and Urban Politics in Istanbul," *City* 18, no. 6 (2014).

15 MAPEG. 2017. "Orta Dönemli Petrol ve Doğal Gaz Arz-Talep Projeksiyonu," http://www.mapeg.gov.tr/petrol/orta%20d%C3%B6nemli%20arz%20 talep%20projeksiyon/Orta_Donemli_Petrol_ve_Gaz_Arz-Talep_Projeksiyonu .pdf.

16 Christopher Houston, "An Anti-History of a Non-People: Kurds, Colonialism, and Nationalism in the History of Anthropology," *Journal of the Royal Anthropological Institute* 15, no. 1 (2009).

17 Three main Kurdish rebellions in this period are the Seyh Sait rebellion in 1925, the Ararat rebellion in 1935, and the Dersim rebellion in 1938.

18 Şemsa Özar, Nesrin Uçarlar, and Osman Aytar, *From Past to Present a Paramilitary Organization in Turkey: Village Guard System* (Diyarbakir Institute for Political and Social Research, 2013).

19 The proverb "The Kurds have no friends but the mountains," resurfaced and gained popularity as a rhetorical device following after US President Trump decided to withdraw US troops from north-eastern Syria in October 2019. For US betrayal and the history of other betrayals, see Michael M Gunter, "The Kurdish Question in Perspective," *World Affairs* 166, no. 4 (2004): 197.

20 H. Akın Ünver, "Schrödinger's Kurds: Transnational Kurdish Geopolitics in the Age of Shifting Borders," *Journal of International Affairs* 69, no. 2 (2016): 65.

21 *Milliyet,* "Doğa Teröristlere Kolaylık Sağlıyor," 11 December 2007.

22 Thongchai Winichakul, *Siam Mapped: A History of the Geo-Body of a Nation* (Honolulu: University of Hawaii, 1997).

23 Last, "Fruit of the Cyclone," 56.

24 Mezzadra and Neilson, *Border as Method,* 17, 36.

25 Casper Bruun Jensen and Atsuro Morita, "Infrastructures as Ontological Experiments," *Engaging Science, Technology, and Society* 1 (2015); Elizabeth A. Povinelli, Mathew Coleman, and Kathryn Yusoff, "An Interview with Elizabeth Povinelli: Geontopower, Biopolitics and the Anthropocene," *Theory, Culture & Society* 34, no. 2–3 (2017).

26 Nazan Üstündağ. "Bakur Rising: Democratic Autonomy in Kurdistan," *ROAR Magazine* 6 (2017), https://roarmag.org/magazine/democratic -autonomy-municipalism-kurdistan.

27 Amberin Zaman, "Kurdish Stamp Commemorating Pope's Visit to Erbil Fans Turkish Conspiracy Fears," *Al-Monitor* (March 10, 2021), https://www .al-monitor.com/originals/2021/03/turkey-lashes-out-erbil-pope-visit-stamp -iraq.html.

28 Also see Umut Yıldırım, "Space, Loss and Resistance: A Haunted Pool-Map in South- Eastern Turkey," *Anthropological Theory* 19, no. 4 (2019): 440–69.

29 İsmail Beşikçi, *International Colony Kurdistan* (London: Taderon Press, 2004). On the genealogy of "international colony," see Deniz Duruiz, "Tracing the Conceptual Genealogy of Kurdistan as International Colony," *Middle East Report* 295 (Summer 2020), https://merip.org/2020/08/tracing-the-conceptual -genealogy-of-kurdistan-as-international-colony.

30 Walid Siti, "A Poem to the Mountain at the Edge of the World" (2019), Walidsiti.com, https://www.walidsiti.com/videos-installations?pgid=k95koq8w -03e90158-d39a-411c-ba46-81fb9d25943d.

31 Quoted from Rose Issa, curator and author of the catalogue of the Leighton House exhibition, London 2008, *Land on Fire,* a retrospective of the works of Walid Siti which covered the period 1997 to 2008. In Lutz Becker, "Walid Siti: New Babylon/Yeni Babil," *Galeri Zilberman* (2014), 2.

32 Walter D. Mignolo, *Local Histories/Global Designs: Coloniality, Subaltern Knowledges, and Border Thinking* (Princeton, NJ: Princeton University Press, 2012).

15 TESTOSTERONE

j.r. latham
kate seear

They did not see me as human at all.

– Caster Semenya, 2019[1]

Most commonly understood as "the male sex hormone," testosterone circulates and acts in almost all human bodies (and many nonhumans). Testosterone's elementality to sex – specifically maleness – is so ideologically strong as to be taken for granted, including within endocrinology (as evidenced by its misleading designation as "male").[2] While many chapters in this book explore an anthropogenic element by outlining its history, analysing a case study or enactment, or investigating how it acts in a political climate, we have taken a different approach. In our chapter, we aim to *resist* the elementality of testosterone.[3] Instead, we draw attention to ways that social practices, structures and institutions make connections between sex and testosterone, and how we come to understand (and be understood as) human (or not) via testosterone. We consider how testosterone is mobilized in relation to human rights to connect bodies and sport, fairness and equity, and to help constitute ways of being human (or less-than-human). Inspired by Annemarie Mol's *The Body Multiple*,[4] we divide this chapter into three parts as a way of representing the strains of living between discourses and worlds, and of resisting and disrupting the constitutive function of testosterone. We recognize the challenge that structuring a chapter in this way may pose, but invite you to read this work however appeals to you: perhaps by taking breaks by paragraphs, perhaps by reading one part at a time in succession, and/or perhaps by reading in one way and then re-turning in another. We hope that in holding that difficulty, and/or in the turning forward and backward of the pages (or scrolling up-and-back or side-to-side), we might convey to you something of the giddiness that multiplicity's instability presents; what it is like to be elementized by testosterone in a way that is … not … quite … "right" …

Doing Human Right

J: I'm always nervous going to new places. Stepping over the threshold into the unknown. Whatever it is, a pizza parlour, a train station, a new city: I breathe staccato. The glass sliding doors shudder as they part for me – a first act of recognition. I am solid, I am here. The humidity hits me immediately; the air is thick, making it even harder to breathe. I clench my jaw and push through the turnstile. A little guy patters out almost knocking his head on my elbow with a fella close behind hissing, "James-don't-run!" A sign reads simply: NO GLASS. NO MOBILES. The walls are that kind of murky white that tries – and fails – to say **clean**.[5] *Always lingering in the air to some degree or another, and here it's faint, which is a good sign, the whiff of urine. The rough concrete floor reminds me of the gravel paths at my suburban primary school. I strip down and place my bag carefully into one of the shimmering lockers – the sheen of condensation dancing with the sunshine coming through opaque*

What Human Rights Do

What are the purposes and effects of human rights? What is it that people think they do? The notion that law reforms should be guided by human rights has increasing prominence within policy talk and academic research alike. In Australia, for example, the federal constitution contains some limited rights protections (including the right to vote and the right to freedom of religion), but Australia has no national bill of rights, and calls for such legislation have been ongoing for decades. But what are human rights, exactly? Some common approaches view rights as a unifying logic for more generous and humane policies and laws. Others see them as a safeguard on state power. But might they function in other, more problematic ways? What does it mean to *have rights* or, to have rights bestowed upon us? Who and how are such choices made? And what does it mean to have these rights violated or otherwise refused? We draw upon ideas from social studies of science, feminist scholarship, and critical human rights literature to argue that human rights processes can sometimes work to instantiate particular and limited notions of "the human." That is, we want to think about how human rights operate in the making of who counts as human.

To begin, let's outline a series of orthodoxies. A typical definition of human rights is that they are literally the rights one has by virtue of being part of the human race. In this sense, as legal scholar Costas Douzinas notes, human rights have come to be seen as totally "synonymous with being human."[6] Modern rights are *clerestory windows – and walk out into the concourse.*

There aren't too many people around. It's after peak hour but before classes start. A large red clock flickers between the time and 32°. I sit on the edge, dip a toe in sheepishly and then push off my hands into the water. I plunge myself down below the surface and then straight back up, shaking my hair like a Labrador at the beach. I adjust my goggles and kick off the wall.

I took swimming lessons as a child – sensibly compulsory in Australia – and I hated it. Often they were at the beach in the early morning when the water felt freezing. I loathe being cold. And the ocean scares

often said to date back to the 1948 *Universal Declaration of Human Rights,* which embodies the widespread international enthusiasm for human rights as a unifying framework for the protection of apparently universal values, including the protection of autonomy, privacy and bodily integrity. As leading human rights scholar Samuel Moyn points out, the discourse of human rights evokes hope, signalling "the possibility of a better life."[7] Human rights are a "recognizably utopian program."[8] A common claim among historians is that the universal declaration was a "natural" and perhaps even "inevitable" result of progress. Some even treat rights declarations as the product of an apparently widespread consciousness of guilt and shame in the wake of the Holocaust. But all of these orthodoxies have been critiqued. We'll just note four of the most important critiques and observations here.

The first critique has to do with the *origins* of rights. Rights did not evolve "naturally"

The Right to Be Human

By looking to the case of South African middle-distance runner Caster Semenya at the Court of Arbitration for Sport (CAS), we can see how processes of assessing and applying human rights are revealing, as they involve the making of connections between sex and bodies, drugs and humans. The case is long and complex, and we have no chance of doing it justice here. Briefly, in 2009 at the age of 18, Semenya won gold at the World Championships in Athletics. Some of her competitors complained to the governing body (the International Association of Athletic Federations [IAAF]; recently re-named "World Athletics") and she was asked to undergo sex verification testing.[9] Those results were not publicly released, however certain data were leaked and the IAAF soon after developed guidelines governing "the eligibility of females with hyperandrogenism."[10] Hyperandrogenism means having higher levels of naturally occurring testosterone, above the "normal" range. Importantly, it is "a medical concept with no analogue in men."[11] The guidelines required *me. Signing up to snorkel at the Great Barrier Reef some years ago, I did not have a simple answer to the question: "Are you an able swimmer?" But I love the water. Some of my fondest childhood memories are of splashing about in a friend's backyard pool or at the shoreline with my sister.*

At first I can hardly keep from drowning. I flail from one end to the other, puffing out of the water at each end to catch my breath. It's been at least 20 years since I've swam a lap, and never have I done so by choice. I can't think of anything but my own breathing: to get air in my lungs – and not water. I remember the strokes. I know them: breast,

but are products of Western thinking, reflecting colonial preoccupations and masculine ideals. As Moyn argues:

> Rather than originating all at once as a set and then merely awaiting later internationalization, the history of the core values subject to protection by rights is one of *construction* rather than discovery and *contingency* rather than necessity.[12]

Thus, rights are not a natural or inevitable manifestation of progress; they were not discovered but *made*. The first critique thus acknowledges that *values* are embedded in human rights. These values are not only humanist and individualist, but reinforce gendered, racialized, and heteronormative ideals. Moyn draws our attention to the way that human rights instantiate a notion of the human subject as free and equal, and bring into question putatively fundamental elements of being "human" – such as autonomy, privacy, and bodily

---

such women to lower their testosterone levels by taking testosterone-suppressing drugs. Semenya challenged the guidelines, claiming (among other things) that the rules violated her human rights. Specifically, she argued that they were discriminatory against women, intersex people, and people with certain innate biological traits. The IAAF, in contrast, argued that any rights limitations within the rules were justifiable on the basis that they were necessary to protect the integrity and sanctity of sport. Here, the integrity and sanctity of sport is connected to another concept: that sporting competitions must be "fair." By this, the IAAF meant that it would be unfair for other women to have to compete against Semenya, given her apparent natural advantages. Even those with a cursory understanding of sport may well feel confused at this point. Isn't sport, at least in part, a celebration of apparently "natural attributes" and "genetic gifts"? Did swimmer Ian Thorpe's competitors complain that his relatively large feet posed an unfair disadvantage to his competitors? Or is it precisely these ostensibly natural advantages that we admire when we witness the extraordinary feats of Olympic success? How is it then, we might ask, that

free, back; I re-enact the movements of some of my favourite Olympians. The memories flood my muscles, spark in my blood vessels, tingle in my skin. I can do this. *My body recalls the motions. I hold my fingers taut and slice through the gentle waves. My joints, so often aching with the pains of the everyday, rotate smoothly. I stretch and heave, pushing and pulling, together–apart, I feel my muscles and lungs strain, blood rushing through my aorta like it hasn't done in months. The water does not resist, it invites; it welcomes me.*

Overwhelmed to the point of paralysis almost constantly on dry land, I can move in the water. Enveloped

integrity. Each of these rights has histories, and futures. Thus, this first critique invites us to question those elements of existence that human rights have isolated and prioritized as if natural.

The second critique has to do with the supposed *universality* of human rights, specifically the taken-for-granted notion that they are rights ascribed to individuals equally and by virtue of their participation in the human race. Rights are in fact not universal, nor enjoyed equally – a point political philosopher Hannah Arendt famously made in her critique of the relationship between rights and citizenship. Reflecting on statelessness, refugees and citizenship in the aftermath of the Second World War, Arendt argued that the very idea of human rights "broke down" in this period, through the realization that rights were not bestowed on all subjects as equals.[13] As Arendt points out, the conferral of rights is instead a *political process* connected to various concerns, including nationalism, religion, race, and so

the biological attributes of some people are viewed as more natural and (thus) acceptable than others? What does this tell us about the way that sport constitutes the human?

The CAS decided in favour of the IAAF. Importantly, the IAAF rules were found to be discriminatory, but the court also found that "such discrimination is a necessary, reasonable and proportionate means of achieving the aim of what is described as the integrity of female athletics."[14] In general terms, the end result is that Semenya, and other women like her, cannot compete in athletics competitions unless they are willing to take testosterone-suppressing drugs. In this, we see a series of important tensions regarding how "nature" is understood in sport. In one respect, the CAS defers to "nature" and "biology," claiming that Semenya's naturally occurring testosterone is the key determinant in her performance – a claim that is "profoundly contested."[15] At the same time, the CAS is saying that some athletes should tamper with their "natural" bodies by taking drugs. This is in stark contrast to the prohibition on doping in sport, the enforcement of which routinely limits athletes' rights to privacy, bodily autonomy, and due

in its silky embrace, I feel ig-
nited; I feel alive. All of the
touches of daily life – the
pleasures, the violations – I
can saw through them; they
wash over me. **Water off a**
duck's back.

I've spent a lot of time
thinking about how I'm af-
fected by testosterone – how
testosterone shapes the world
around me. I've read books
about it.[17] *And written*
one.[18] *I've researched how*
men who inject testosterone
make sense of illicit prac-
tices as normative impulses
of masculinity, in which
work on the self is made to
appear natural.[19] *Some of*
these men gave beautiful,
complex accounts of their
careful regimes of living in
the world; how they worked

on. Thus, if one must "be human" to have
rights and not all subjects have them, then
"not all subjects can be said to be human."[16]

This leads us to the third observation
about rights, which concerns the *procedural
nature* of rights bestowal. Although we
might believe that we have human rights
absolutely, they can be limited in certain
circumstances. This might happen where
a government wants to propose a law that
would limit one's rights – the right to freedom of movement, for example. In deciding whether rights should be limited,
legal tests are used. A common test might
say that it is appropriate for the proposed
law to limit rights if the law would achieve
a "legitimate objective" and if it is a *reasonable and proportionate* means of achieving
that objective, like staying at home to stop
the spread of a pandemic disease, for instance. As legal scholar Kyle McGee points
out, doctrinal constructs such as reasonableness, proportionality, and legitimacy are
not self-executing or self-applying. Rather

---

process on the basis – or so we are told – that only "natural" bodies have a
place in sport. Certain kinds of drug use are prohibited precisely because
it would be both unfair and counter to the spirit of sport to permit bodies
that are "unnaturally enhanced." This conceptual confusion in turn raises
important questions about sporting integrity and notions of fairness in
sport. How is it fair to require some athletes to *take* performance-mod-
ifying drugs in order to compete? Is the integrity of sport undermined
when some aspects of natural bodies are positioned as inherent to elite
sport, and to be celebrated, while others are deemed mistakes to be corrected? In Semenya's case, bizarrely, drug use in sport is positioned not as
problematic, but as positive.[20] As bioethicist Katrina Karkazis and feminist
science scholar Rebecca Jordan-Young contend,

> In a context in which T[estosterone] alone is deemed to determine
> advantage and disadvantage, what makes sense and is valued as legit-
> imate in this scene is the sense of injustice expressed acutely by the
> women who did not win the race. But women investigated for possible

*to do better, to **be** better –
men.*

*When I first injected tes-
tosterone, it made me sick.
I'd never understood the
word **penetrate** until then. I
could feel every cell that was
being broken through – the
skin, the fat, the muscles –
torn apart; violated. It was
like that first nervous entering of the icy waters of my
childhood; I was reluctant
and slow – I hesitate, work
up the resolve and grimace
as I push the needle down
into my thigh. I never got
used to it. I would break out
in a nervous sweat, and
have to sit on the floor to
try to de-escalate my panic
afterwards. Why-why-why!?
They demand.*[24] *And soon
enough: What-happened-*

they must be manipulated, balanced, and
coordinated: the "doctrinal category, figure
or rule does not exist except as a function
of its application *which varies every time.*"[21]
These tests frequently turn "on the court's
perception of what is customary or generally expected by participants, or of what
'everyone knows' or 'should know,'" as
lawyers argue it and courts interpret it in
any given scenario.[22] The meaning of all of
these things, then, differs between contexts.
Philosopher Bruno Latour suggests that this
tells us something about the law itself – that
its "modes of connecting" are constitutive.[23]
That is, law is synonymous with connections that are articulated in its name. Thus,
whether something is "reasonable" depends
on how meanings emerge in court, for example, how concepts such as freedom and
privacy and bodily integrity are connected.
The essence of human rights is thus proces-
sual, and lies in *practices* such as these tests.

The fourth point concerns the *normative,
constitutive function* of rights. In this respect,

---

high T face harms that are nowhere in the picture: having their identity publicly questioned, their genitals scrutinized, the most private
details of their lives subject to "assessment" for masculinity, their ca-
reers and livelihoods threatened, and being subject to pressure for
medically unnecessary interventions with lifelong consequences.[25]

How are connections being made in this case that fail to address these harms?
The CAS decision shows how rights can be both acknowledged and discarded
with the same breath and how fundamental normative conceptions of the
human are to those very processes. In these deliberations, "non-normative"
subjects are not acceptable as we *are*: we are acceptable only to the extent that
we are able and prepared to conform to specific norms. In all of this, human
rights are not a bulwark or obstacle to this normalization, but central to it.
What we mean by this is that the decision to exclude Semenya stands *not despite human rights, but because of them* – because of the way the human emerges
through the very practices of interpreting human rights. In this case, human
rights have been deployed in ways that pit Semenya against the majority and
then undermine her, not just as a woman but as a human being. She has

*what-happened-what-hap-
pened!? What did it **do**?*

*I work every day to both
undercut and expand what
it means to be a man, in-
cluding at the pool. I am a
different kind of white man,
and that's visible, whether
(some) people can identify
the host of markers I carry
with me or not. Did testosterone make it possible for me
to walk through the natato-
rium in nothing but a little
pair of shorts and flip-flops?
Hardly. A salary, an in-
jury, a friend, the desperate
pressure to finish my book
and what it all means –
these were all involved,
and so much more. **Pass-
ing** is such an insincere,
inadequate term. Whether
I'm "passing as a man" or*

Douzinas notes that perhaps "the greatest achievement of rights is ontological:
rights contribute to the creation of human
identity."[26] For instance, the recognition
of rights for a category of persons called
"women" helps to constitute the subject
position of women. This in turn functions
to constitute the very contours of the wom-
an-as-human because, as Douzinas puts it,

> an individual recognized as a legal subject
> in relation to women's rights is *accepted as
> the bearer of certain attributes and the benefi-
> ciary of certain activities* and, at the same
> time, as a person of a particular identity
> which partakes amongst others of the *dig-
> nity of human nature.*[27]

Building on Arendt's critique, cultural
studies scholar John Erni notes that human rights thus possess "the capacity to
render the 'subjects' of rights as differently
human."[28] For example, the process of bal-
ancing rights against one another, or of

rights, but they are rights that hold only if she is prepared – and this is the key
point – to be *differently human. **What is fair about that?***

The sporting attachment to a principle of fairness produces great enthusiasm about the capacity of sport to foster and promote rights, and there is
indeed a long history of iconic moments in sport that have furthered rights.
This enthusiasm is not misplaced, but we must also be cautious about the enormous power of sporting organizations to interpret, apply and balance rights in
ways that mark some subjects as human, and others as less-than-human. Most
sport adopts a binary, sex-segregated competition model and then marks as
a problem those humans that do not fit its own definitions of sex. Here, the
"problem" thrown up by sex-segregated sport is one that ends up attaching –
not to the sporting bodies that insist on binary categories but – to the bodies
of those who do not fit the categories that sport defines for itself. It is vital that
we interrogate the logic that sporting bodies use to justify these approaches,
especially where they result in the exclusion of some by mobilizing vague and
ill-defined concepts such as fairness, or the spirit, integrity and sanctity of sport.

If human rights get made via modes of connecting sex, drugs, rights
and bodies, then it is worth considering how they might be articulated and

"passing as not trans" isn't relevant to me. What is relevant is if I'm being treated with respect and dignity; if I'm being regarded as another human among my own kind.

*In so many contexts these things are indistinguishable. To be read as **not** a white man, to be read as **unclear** in regards to sex is to suffer the burdens of social ostracism in myriad insidious forms. Those of us not clearly, boringly, adhering to social norms are punished. At my pool, they work hard to make another world possible. A world where all kinds of people can strip down and enjoy the facilities. For me, it is life-changing.*

testing rights against apparently self-evident standards (such as reasonableness) renders some rights, subjects, or ways of being as less valid, which may in turn subject them to being "corrected." To take just one example, the forced containment and treatment of people who use drugs has been historically justified on the basis that it is being done to help them regain their "capacity" and become more "autonomous."[29] In this way, the ostensibly "legitimate objective" of forced treatment is shaped by deeply entrenched, normative ideals of what it means to be "properly" or "appropriately" human. Rights function in this example to authorize medico-legal interventions, governance, violence and control in a bid to "humanize" subjects. In other words, rights constitute the human, and they do so by enjoining appropriate modes of being human. Human rights are thus a fundamental element in the constitution of the subject, rendering some subjects as human, and others as *less than fully human.*

---

practised in sport differently. What if fairness in sport was envisaged not as being about how unfair it is to have to compete against those deemed inappropriately "female," but about how unfair it is to subject athletes to such inhumane scrutiny? What if we thought about the spirit of sport as one in which different ways of living and being were celebrated and endorsed – as they are for Michael Phelps, the swimmer whose proportionately large wingspan is seen as a gift to be celebrated rather than an impairment to be corrected? What if the integrity of sport was considered instead to be about the respectful acknowledgment of human diversity? What if the sanctity of sport was similarly reimagined: to be not laden by sexist assumptions about the apparent hazards of different bodies, but by openness to their affordances? How might we enjoin the excitement at what different bodies can do that is inherent to sports to *include* athletes like Semenya? These are the challenges that lie before sporting organizations moving forward. They require a radical rethink of the foundational assumptions of sport, and the remaking of links between sex, drugs, bodies, and rights. In so doing, we might find new and more generous ways of thinking, playing, and being human.

NOTES

1 [Mokgadi] Caster Semenya, "I Wanted to Be a Soldier," *The Players' Tribune*, 27 September 2019, http://projects.theplayerstribune.com/caster-semenya -gender-rights.

2 See Nelly Oudshoorn, *Beyond the Natural Body: An Archaeology of Sex Hormones* (London and New York: Routledge, 2005 [1994]).

3 We note that sperm and testosterone have earned their own chapters as anthropogenic elements, while ova, oestrogen and progesterone have not.

4 Annemarie Mol, *The Body Multiple: Ontology in Medical Practice* (Durham, NC, and London: Duke University Press, 2003), ix.

5 See Lucas Crawford, "Derivative Plumbing: Redesigning Washrooms, Bodies, and Trans Affects in Ds+r's Brasserie," *Journal of Homosexuality* 61, no. 5 (2014): 621–35.

6 Costas Douzinas, "Justice and Human Rights in Postmodernity," in *Understanding Human Rights*, eds. C.A. Gearty and Adam Tomkins (Mansell, 1996), 123.

7 Samuel Moyn, *The Last Utopia: Human Rights in History* (Cambridge, MA: Belknap Press, 2012), 1.

8 Moyn, *The Last Utopia*, 1.

9 See Tavia Nyong'o, "The Unforgivable Transgression of Being Caster Semenya," *Women & Performance: A Journal of Feminist Theory* 20, no. 1 (2010): 95–100; Shari L. Dworkin, Amanda Lock Swarr, and Cheryl Cooky, "(In)Justice in Sport: The Treatment of South African Track Star Caster Semenya," *Feminist Studies* 39, no. 1 (2013): 40–69.

10 These 2011 guidelines were subsequently revised in 2018 to restrict participation specifically in middle-distance running events.

11 Katrina Karkazis and Rebecca M. Jordan-Young, "The Powers of Testosterone: Obscuring Race and Regional Bias in the Regulation of Women Athletes," *Feminist Formations* 30, no. 2 (2018): 19.

12 Moyn, *The Last Utopia*, 20; emphasis added.

13 Hannah Arendt, *The Human Condition* (Chicago: The University of Chicago Press, 1958), 297.

14 Semenya, Mokgadi Caster v. International Association of Athletics Federations [CAS 2018/O/5794] & Athletics South Africa v. International Association of Athletics Federations [CAS 2018/O/5798], Court of Arbitration for Sport, 2018, https://www.tas-cas.org/en/jurisprudence/recent-decisions/article /cas-2018o5794-mokgadi-caster-semenya-v-international-association-of-athletics -federations-cas-2.html, 160.

15 Katrina Karkazis and Morgan Carpenter, "Impossible 'Choices': The Inherent Harms of Regulating Women's Testosterone in Sport," *Journal of Bioethical Inquiry* 15, no. 4 (2018): 580.

16 Kate Seear, *Law, Drugs and the Making of Addiction: Just Habits* (London: Routledge, 2020), 42.

17 Werner Reiterer, *Positive: An Australian Olympian Reveals the Inside Story of Drugs and Sport* (Sydney: Pan Macmillan Australia, 2000); Paul B. Preciado, *Testo Junkie: Sex, Drugs, and Biopolitics in the Pharmacopornographic Era*, trans. Bruce Benderson (New York: The Feminist Press, 2013); Cordelia Fine, *Testosterone Rex: Myths of Sex, Science, and Society* (New York: W.W. Norton & Company, 2017); Rebecca M. Jordan-Young and Katrina Karkazis, *Testosterone: An Unauthorized Biography* (Cambridge, MA: Harvard University Press, 2019).

18 J.R. Latham, *Making Maleness: Trans Men and the Politics of Medicine* (University of Minnesota Press, forthcoming).

19 J.R. Latham, Suzanne Fraser, Renae Fomiatti, David Moore, Kate Seear, and Campbell Aitken, "Men's Performance and Image-Enhancing Drug Use as Self-Transformation: Working Out in Makeover Culture," *Australian Feminist Studies* 34, no. 100 (2019): 149–64.

20 For analysis of the harms of these interventions, see Karkazis and Carpenter, "Impossible 'Choices.'"

21 Kyle McGee, "On Devices and Logics of Legal Sense: Toward Socio-technical Legal Analysis," in *Latour and the Passage of Law*, ed. Kyle McGee (Edinburgh: Edinburgh University Press, 2015), 61–92, 69; emphasis added.

22 McGee, "On Devices and Logics of Legal Sense," 69.

23 See Bruno Latour, *The Making of Law: An Ethnography of the Conseil d'Etat* (Cambridge and Malden: Polity Press, 2009).

24 See J.R. Latham, "(Re)Making Sex: A Praxiography of the Gender Clinic," *Feminist Theory* 18, no. 2 (2017): 177–204.

25 Karkazis and Jordan-Young, "The Powers of Testosterone," 28.

26 Costas Douzinas, *Human Rights and Empire: The Political Philosophy of Cosmopolitanism* (London and New York: Routledge-Cavendish, 2007), 7.

27 Douzinas, "Justice and Human Rights in Postmodernity," 127; emphasis added.

28 John Erni, *Law and Cultural Studies: A Critical Rearticulation of Human Rights* (Abingdon and New York: Routledge, 2019), 37.

29 Seear, *Law, Drugs and the Making of Addiction*.

16 VIRUS

frederic keck

In the French sociological tradition, the concept of "the element" plays a major role in connecting infrastructures and ontologies, or in more ancient terms, social organizations and collective representations. Think of the titles of the major works of two of the most significant European thinkers of the social: Emile Durkheim's *Elementary Forms of Religious Life* (1915) and Claude Lévi-Strauss's *Elementary Structures of Kinship* (1969). The element, in the country of René Descartes, is a logical piece that can be assembled with other pieces, after scepticism and doubt have deconstructed all human perceptions. The ethnographic encounter is taken by French anthropologists as an opportunity to deconstruct what was taken as evident and reconstruct thinking on the basis of European logic. But after Rousseau and the French Revolution, such a process of deconstruction and reconstruction is also highly political, as the ethnographic encounter became an opportunity to rebuild the social contract on the basis of justice and equity. The "elementary" is the "archaic," not in the chronological sense of evolutionary anthropology but in a logical, moral, and political sense. The elementary constraints other elements to bind together in a collective entity.

This helps explain the massive presence of the notion of "element" in Marcel Mauss's essay on the gift, first published in 1923 in Durkheim's journal, *Année sociologique*. In this perspective, the most surprising quotation in this famous essay is: "The *res*, prestation or thing, is an essential element of the contract."[1] This means, in one sense, that there is no "thing" without a social contract that defines to whom this thing should belong. But it also means that there is no social contract without "elements," that is, things that can be exchanged. In Mauss's analysis, the simple act of giving and receiving entails a collective representation which binds together those who take part in the exchange. And this collective representation is highly ambivalent, since it builds social order on a potential violence; gift-giving relations are agonistic as they involve competition in a regulated form. This explains that the *potlatch*, described by Franz Boas in the Northwest Coast, or the *kula*, analysed by Bronislaw Malinowski in the Trobriand Islands, are conceptualized by Mauss as "total social facts." In each, the exchange of shells, clothes, or masks becomes the opportunity to display social divisions and produce prestige, thus connecting economy and culture. The table of elements is dynamic as elements are constantly circulating to produce value in social life.

How can this view of elements help us conceptualize the contemporary situation of the COVID-19 pandemic? Can we consider viruses as elements in the French sociological sense of the term? Viruses are pieces of information that seek to replicate in the nucleus of biological cells, and they are often described as "quasispecies" or "clouds" because they act as unbounded collectives exerting evolutionary pressure on the cells in which they replicate.[2] But the way they circulate – their virality – and the things in which they are captured – microscope images, viral banks, vaccines, masks, and so on – have a lot of impact on the way we want to conceptualize the social contract. Can we suppose that viruses are gifts in the sense, articulated by Mauss, that they are things exchanged and regulated? Can we consider the pandemic as a "total social fact," in which the elementary structures of contagion combine to produce, through the complexities of contemporary apparatuses of biopower, a new social contract, one that emerges after the sceptical doubt of the epidemic crisis and the diabolic moment of collective panic? By examining three techniques of pandemic preparedness – sentinels, simulation, and stockpiling – I suggest that viruses fit the Maussian grid of the "total social fact," displaying the elements of social life.[3]

As global regulations attempt to control the emergence of viruses, pandemics are tracked in their origins. In the last 40 years, global alerts about pandemics have raised public awareness that new infectious diseases spread because of the ecological and economic transformations that amplify the mutations of viruses in animal reservoirs. The Ebola and Lassa viruses emerged after being transmitted from bats to monkeys to humans. HIV was transmitted from non-human primates to humans.[4] Bovine spongiform encephalopathy was transmitted from sheep to cows to humans.[5] Influenzas have been transmitted from birds to pigs to humans. SARS-like coronaviruses have been transmitted from bats to humans via civet cats, camels, and potentially pangolins.[6] As a result, there has been increasing collaboration between veterinarians and physicians to control zoonotic diseases, under the "One Health" agenda.[7]

Different kinds of maps have been produced by health professionals to localize the origins of zoonotic viruses. Global maps attempt to define hotspots in which the proximity between human and non-human animals multiplies the chances of transmission of viral mutations, such as bushmeat markets in Africa or wet markets in China.[8] Diagrams are built to visualize the chains of transmission between animal reservoirs, such as bats or birds, and humans, considered in these analyses as "epidemiological dead ends."[9] These maps have the schematic power of enticing health authorities to act on the origins of pandemics, by regulating animal markets or eradicating animal reservoirs. Following the model of the "patient zero" in contact with a wild untamed origin of contagion, they raise an ambivalent mix of curiosity and fear.

Virologist Nathan Wolfe and geographer Jared Diamond have built a successful narrative of the origins of pandemics.[10] After starting a career as a zoologist at Oxford University, Wolfe realized that the study of viruses provides opportunities to "discover new species and catalog them" in the manner of a nineteenth-century naturalist.[11] He then set up a private company, Global Viral Forecasting Initiative, whose goal was "to hunt down these events, the first moments at the birth of a new pandemic, and understand them to stop them before they reach a global level." Defining himself as a "virus hunter," Wolfe is fascinated by the idea that viruses are particularly transmitted by hunting practices, either those of humans who hunt non-human primates for bushmeat, or those of non-human primates themselves, since, he argues, hunting emerged in our shared archaic ancestors. Wolfe describes hunting as an intimate relation between two living beings in which one spends a lot of energy to reproduce at the expense of the other: "From the perspective of a microbe, hunting and butchering represent the ultimate intimacy, the connection between one species and all the various tissues of another."[12] Wolfe can take the perspective of viruses, like Amazonian shamans, because viruses are relational entities that need to infect living organisms and connect different species and milieus. Building a bank of viruses thus offers him the possibility to anticipate the transformation of ecological niches by looking at viral genetic structures. Like other virus hunters, Wolfe builds phylogenetic trees that display the kinship relationships between viral strains in order to anticipate the next pandemic virus. Just as the table of elements leaves empty spaces for the discovery of new chemical components, phylogenetic trees have branches that drift apart, indicating a viral mutation among other animal species that might spill over to humans.

Jared Diamond, who started his career as an ornithologist in Papua New Guinea, tells the same story on a larger scale: the catastrophic encounters between civilizations, which he famously describes as "collapses," are due to viral infections. Diamond's main example is the violent collision between European colonizers and Amerindian societies. Because they had no immunity to the diseases European societies had co-evolved with through centuries of contact with Asian societies, Amerindian societies collapsed under the encounter with, among other diseases, smallpox and tuberculosis.[13] For Diamond, this collapse replicates the diseases that probably affected earlier human societies when they shifted from hunting to pastoral regimes during the Neolithic revolution, as the animals transmitted pathogens to the humans with which they shared the domestic space. "The lethal gift of livestock," in Diamond's word, is understandable in a Maussian sense of poison and present, or as the "accursed share" of domestication in the sense of Bataille.[14] Humans give care to domesticated animals and receive in return goods such as leather, milk, or meat. But if the conditions of domestication change too dramatically, then the goods they receive may turn into viruses. Thus, for Diamond, the twentieth century "livestock revolution" has similar dramatic consequences to the Neolithic revolution: the abundance of zoonotic viruses. Societies can slow down the circulation of these "gifts" only if they learn to live with them.

While social anthropology and post-modern critical theory have taught us to be wary against essentializing arguments brought up by Wolf and Diamond, their long-term narratives about species contacts, conflicts, and distances across time and space raises an important question: What does the speculation on the origin of zoonotic viruses tell us about our kinship relations with them?[15] This is the main question for the pathology of emerging viruses, such as avian influenza or bat coronaviruses. Malik Peiris, head of the Pasteur Institute of the University of Hong Kong, replied to this question with the hypothesis of the "cytokine storm": viruses coming from distant species trigger an inflammatory response in the host because the immune system does not recognize its antigenic information.[16] The failure of the immune system is explained by the capacity of zoonotic viruses to lure dendritic cells, whose role is to capture viruses' antigenic information, and present it to other immune cells in order to trigger an adapted immune response. These cells are also called "sentinel cells" because they are situated at the outpost of the immune system and watch over the arrival of unknown pathogens. The encounter between a sentinel cell and a zoonotic pathogen can be considered elementary since it produces the antigens that the immune system needs to encounter future viruses. Virologists with whom I worked at the Pasteur Institute of the University of Hong Kong thus studied the spike protein of coronaviruses to understand how it could bind to the receptor of respiratory cells. They assumed that some viruses could use decoy receptors to inhibit the production of cytokines and bypass the immune response of the organism they invade. The table of elements is a misleading representation for the immunological self: rather than a set of antigens matching the set of potential viruses, we should conceive of it as a search engine for the adequate antigen which can be lured and disrupted.[17] The relation between virus and immune cell is sign communication rather than elementary matching.

The concept of sentinel is pervasive throughout pandemic preparedness today, where it is used to define populations or living beings equipped to detect early warning signals of pandemics. Sentinels take the form of bushmeat hunters carrying Ebola or Lassa viruses in Cameroon, *brigadistas* inspecting water ponds for mosquitoes carrying dengue in Nicaragua, unvaccinated chickens who die first when avian influenza enters a farm in Hong Kong, or hens that seroconvert to West Nile when bitten by mosquitoes in Australia.[18] The implementation of sentinel devices in a given territory involves knowledge of relations between humans, animals, and microbes in that territory, since it aims to detect the early warning signals of ecosystem disruption caused by pathogenic chains of transmission. While hotspots maps, zoonotic diagrams, and images of patients zero trace strong separations between humans and animals, with the project of eradicating the wild from the chains of transmission, sentinel devices finely attune relations between human and non-human animals to mitigate the catastrophic consequences of their disruption.

I have argued that sentinel devices can be compared to myths in the treatment of the origins of pandemics. Claude Lévi-Strauss has shown that the function of myths is to mitigate a tension in relations between human and non-human animals by telling the story of a time when they were not separated, in order to institute these relations at "a good distance."[19] Similarly, sentinels rely on stories about the evolution of species that produce immunological selves in distances between species. I now want to argue that simulations of pandemics play a similar role to rituals, as analysed by Lévi-Strauss and others, in that they reduce the multiplicity of relations between living beings displayed by sentinels to a linear narrative of origin, in order to instantiate order after the disorder caused by a pandemic. There are two uses of simulation applied to pandemic viruses: on computers and on "real ground" (or, *in silico* and *in vivo*). Both iterations aim to perform the reality of viruses so that humans act in a coordinated way.

Computer simulations of viruses rely on mathematical models that trace origins through mechanisms called "molecular clocks." Based on the similarities between viral strains downloaded on GenBank, the National Centre for Biotechnology Information database, they build scenarios of descent to isolate a common ancestor who may be considered as responsible for a pandemic. They also model the mutations that allow viruses to cross species barriers, such as binding receptors or cleavage sites at the surface of the viral capsid. For instance, in 2009, virologists at the University of Hong Kong showed that a virus very close to the pandemic H1N1 virus circulated in Hong Kong pigs as early as 2004. However, this didn't mean, in their view, that Hong Kong pigs should be blamed for the emergence of the pandemic influenza virus.[20] Rather, they argued for a better surveillance of industrial pig farms in North America to detect the early warning signals of the emergence of influenza viruses. The use of viral banks allowed virologists, in that case, to simulate what could happen if emerging viruses were stopped in their animal reservoirs, where they take elementary forms.

Another virus hunter's story shows how blame and blaming orients the quest for the origins of viruses. In 2012, Ron Fouchier and Yoshi Kawaoka, two major US-based virological experts, announced that they had modified the H1N1 virus to make it as lethal to humans as the H5N1, through a series of experiments that resulted in "gain-of-function."[21] Their publications were criticized as introducing information that could be used for bioterrorism, meaning the malicious release or dissemination of biological agents, and they consequently imposed a moratorium on "gain-of-function" on potentially pandemic pathogens. When the H7N9 virus emerged in Shanghai, they showed that "gain-of-function" experiments allowed them to prepare for the pandemic transmission of this new virus, because it provided the target of the few nucleotides necessary to trigger a pandemic, and hunt down these nucleotides in China's food markets. Here again the virus is not an element that could be combined in an experimental box but a sign of what could emerge in natural processes.

By contrast with computer simulations, where action is applied to virtual data, simulations of pandemics on "real ground" distribute capacities to act in order to mitigate reactions of panic usually observed in situations of disasters. Based on scenarios written by disaster experts, these *in vivo* simulations rely on past epidemics to imagine the consequences of future epidemics. Localized in different sites of epidemic transmission (markets, borders, hospitals, airports, community housing, and elsewhere), they pretend as if an emerging virus was really there to regulate the actions of humans (farmers, retailers, nurses, doctors, patients, and others) based on imagined symptoms. Non-human animals are also involved in these simulations, such as chickens in a market or dogs in community housing who can be integrated in the work of care as companion species or destroyed as animal reservoirs. "The aim of such testing through exercises and drills is to assess preparedness; that is, to evaluate performance, ensure compliance, maintain vigilance and improve readiness," writes medical anthropologist Carlo Caduff. "Simulated events allow public health professionals to reveal potential vulnerabilities and address gaps in the preparedness and response plan."[22] Simulations can be situated between ritual and performance. They play on the relational asymmetries between actors of a pandemic to mitigate the tension between species revealed by pandemic viruses. They distribute responsibility, merit, and blame for the management of scarce resources in institutions made vulnerable by pandemics. While sentinels told stories of gift-giving relationships at the origins of the domestication of animals, simulations tell stories of debt and fault at the origins of a singular pandemic virus. While virologists open the narrative of pandemics to these interspecies relationships, public health authorities often restrict them to human institutions. This restriction is more visible in discussions on stockpiling.

The global implementation of national stockpiles for pandemic preparedness can be dated back to the 1976 swine flu scare, which immediately led the US federal government to stockpile vaccines for the next pandemic.[23] This measure was globalized and strengthened, particularly in 2005 in the aftermath of the SARS crisis, when the World Health Organization published the International Health Regulation. This Regulation recommended that member states stockpile vaccines, antivirals, and masks against the next flu pandemic which, at the time, was imagined as probably coming from an avian influenza virus coming from China. After 2005, for instance, Vietnam became a model stockpiling state in its attempt to produce and distribute vaccines for chickens and humans as a measure to prepare for a bird flu pandemic, while Taiwan invested massively in its pharmaceutical industry to compensate for its non-representation at the World Health Organization.[24] But during the COVID-19 pandemic, where no vaccine or antiviral drug was immediately available, stockpiling strategies led to accusations against national states for failing to provide masks to their populations, and to an aggressive "mask diplomacy" whereby China, Taiwan, and Vietnam competed to provide European states with cheap plastic masks. Such generosity was often perceived as a humiliation by former colonial states.

In many respects, the stockpiling of vaccines, antivirals, and masks can be conceived as a "total social fact" as conceptualized by Mauss. Since Louis Pasteur, we know that vaccines are attenuated viruses which trigger an immune memory in the organisms where they are injected, and that fundamental research on viruses in laboratories is accompanied by technological progress in the pharmaceutical industry, where in-novations in the molecular manipulations of vaccines have multiplied over the past century.[25] Research on flu vaccines played a crucial role in the knowledge about the virus itself at a time when electronic im-ages of viruses were unavailable. The structure of the influenza vaccine – with its hemagglutinin and neuraminidase proteins providing the ini-tial letters H and N for the classification of flu viruses – was discovered when chicken embryos were infected with the flu virus, triggering a red stain called hemagglutination.[26] Still today, the production of flu vac-cines relies on a massive industry of chicken eggs (see figure 16.1).[27] We can reasonably argue that vaccines are "gifts" in the Maussian sense of "offer" and "poison": when they are controlled and regulated by the pharmaceutical industry and by national states, they gain value and be-come commodities which can be stockpiled and exchanged. However, we tend to forget the animal origins of vaccines when we discuss the technological performances of pharmaceutical industries, such as the messenger-RNA technique which accelerated the production of vaccines against COVID-19.

The global growth in the production and stockpiling of masks raises other issues. While European surgeons started to wear face mask at the end of the nineteenth century to protect their patients from droplets, and while the use of surgical masks in the public space was popularized by the Cambridge-trained Malaysian physician Wu Lian-Te during the pneumonic plague pandemic in Manchuria in 1910, leading to massive wearing of masks in the US during the 1918 flu pandemic, a major his-torical shift occurred when reusable masks, made of cotton or paper, were replaced by disposable masks made in plastic in the 1960s.[28] The consumption of throw-away or disposable surgical masks increased with the multiplication of surgical operations and the need to stockpile masks for pandemics. A mask, in Mauss's analysis, is the sign that a person is in-complete by itself but must communicate with another person in a chain of gift-giving relationships.

Two recent pandemics – H1N1 in 2009 and COVID-19 – can thus be described as *potlatch* events in the sense of these ritual ceremonies which are described by Boas and explored in depth by Mauss as an example of agonistic gift-giving relationships. By stockpiling vaccines, antivirals and masks, national states, allied with pharmaceutical companies, dis-play their power in a global competition regulated by the World Health Organization with the goal to show other nation states that they care for the health of their populations. As Boas observed a century ago, we are typically in a situation where the rituals of hunting societies simulat-ing invisible entities coming from animals are distorted by the need of

Figure 16.1. Chicken eggs being inoculated with influenza virus (credit: Carlo Caduff)

capitalist societies, thus leading to a strange combination of creation, destruction, and waste.

Lévi-Strauss shows that myths begin with cosmopolitical narratives about a time when humans and animals were not separated and end with police novels about a transgressive crime affecting a society.[29] In the same sense, pandemics start with narratives where viruses reveal a complex web of relations between human and non-human animals and end with controversies about who should be blamed for failing to stock-pile masks, what kinds of drugs should be tested and promoted and how vaccines should be fairly distributed. In these controversies, it is often forgotten that viruses are gifts sent by animals to humans in return for the care they receive – or not.

In a famous 1951 article, Lévi-Strauss compares language and society to suggest that linguists and anthropologists should build an inventory of elementary structures "comparable to the table of elements that modern chemistry owes to Mendeleev."[30] I have suggested that this linguistic met-aphor should be taken seriously by anthropologists and other analysts of social life. I do not mean this in the sense that anthropology should be founded on linguistics, as biology was founded on chemistry thanks to Mendeleev, but rather this metaphor allows us to think of the dynamics of social life through an attention to the mode of existence of signs. Viruses, like signs, circulate between living beings and trigger all kinds of immune response which, taken together, shape the "total social fact" of a pandemic. Viruses can be compared by their size, which determines their capacity to mutate while replicating in biological material, and these mutations are signs that they can connect a range of biological organisms. Just like Mendeleev's Periodic Table of Chemical Elements, the classification of viruses is a model for social anthropology not only because it stabilizes the variety of forms in a framework, but also because it engenders the dynamics of all social forms from the simplest forms of life. Attention to the diversity of these signs is paid by virus hunters rather than by public health authorities who manage pandemics in a pastoral way as threats to mitigate. In that sense, the age of pandemic viruses can be described as realizing Lévi-Strauss's dream of a reiteration of the modes of thinking of hunting societies within the pastoral forms of power.

NOTES

1 Marcel Mauss, *The Gift*, expanded edition, trans. Jane Guyer (Chicago, IL: HAU Books, 2016), 152.

2 E Domingo et al., "Viruses as Quasispecies: Biological Implications," in *Quasispecies: Concept and Implications for Virology* (Springer, 2006); Celia Lowe, "Viral Clouds: Becoming H5N1 in Indonesia," *Cultural Anthropology* 25, no. 4 (2010).

3 See Andrew Lakoff, *Unprepared: Global Health in a Time of Emergency* (Oakland, CA: University of California Press, 2017); Frédéric Keck, *Avian Reservoirs: Virus Hunters and Birdwatchers in Chinese Sentinel Posts* (Durham, NC: Duke University Press, 2020).

4 Robert Gallo, *Virus Hunting: AIDS, Cancer, and the Human Retrovirus* (New York, NY: Basic Books, 1991).

5 Albert Osterhaus, "Catastrophes after Crossing Species Barriers," *Philosophical Transactions of the Royal Society of London* 356, no. 1410 (2001).

6 Yi Guan et al., "Isolation and Characterization of Viruses Related to the SARS Coronavirus from Animals in Southern China," *Science* 302, no. 5643 (2003); Lin-fa Wang and Christopher Cowled, *Bats and Viruses: A New Frontier of Emerging Infectious Diseases* (John Wiley & Sons, 2015).

7 Steve Hinchliffe, "More than One World, More than One Health: Reconfiguring Interspecies Health," *Social Science & Medicine* 129 (2015).

8   Hannah Brown and Ann H. Kelly, "Material Proximities and Hotspots: Toward an Anthropology of Viral Hemorrhagic Fevers," *Medical Anthropology Quarterly* 28, no. 2 (2014): 280–303.

9   Christos Lynteris, "Zoonotic Diagrams: Mastering and Unsettling Human-Animal Relations," *Journal of the Royal Anthropological Institute* 23, no. 3 (2017).

10 Nathan Wolfe, Claire Panosian Dunavan, and Jared Diamond, "Origins of Major Human Infectious Diseases," *Nature* 447, no. 7142 (2007).

11 Nathan Wolfe, *The Viral Storm: The Dawn of a New Pandemic Age* (New York, NY: St. Martin's Press, 2011), 3.

12 Nathan Wolfe, *The Viral Storm: The Dawn of a New Pandemic Age* (New York: St. Martin's Griffin, 2012), 48.

13 Jared M. Diamond, *Guns, Germs and Steel: A Short History of Everybody for the Last 13,000 Years* (Random House, 1998).

14 Georges Bataille, *The Accursed Share: An Essay on General Economy*, trans. Robert Hurley (Zone Books, 1988).

15 Donna Haraway, *Staying with the Trouble: Making Kin in the Chthulucene* (Duke University Press, 2016).

16 Malik Peiris et al., "The Role of Influenza Virus Gene Constellation and Viral Morphology on Cytokine Induction, Pathogenesis, and Viral Virulence," *Hong Kong Medical Journal* 15, no. 4 (2009).

17 David Napier, *The Age of Immunology: Conceiving a Future in an Alienating World* (Chicago: University of Chicago Press, 2003).

18 Alex Nading, *Mosquito Trails: Ecology, Health, and the Politics of Entanglement* (Oakland, California: University of California Press, 2014); Peter Doherty, *Sentinel Chickens: What Birds Tell Us about Our Health and the World* (Melbourne: Melbourne University Press, 2012).

19 Claude Lévi-Strauss, *Anthropology and Myth: Lectures 1951–1982*, trans. Roy Willis (Hoboken, NJ: Blackwell, 1987).

20 Gavin Smith et al., "Origins and Evolutionary Genomics of the 2009 Swine-origin H1N1 Influenza A Epidemic," *Nature* 459 (2009): 1122–5.

21 Ron A.M. Fouchier et al., "Gain-of-Function Experiments on H7N9," *Science* 341, 6146 (2013): 612–13.

22 Carlo Caduff, *The Pandemic Perhaps: Dramatic Events in a Public Culture of Danger* (Oakland, CA: University of California Press, 2015), 130.

23 Edwin D. Kilbourne, "Influenza Pandemics of the 20th Century," *Emerging Infectious Diseases* 12, no. 1 (2006).

24 Natalie Porter, *Viral Economies: Bird Flu Experiments in Vietnam* (Chicago, IL: University of Chicago Press, 2019).

25 Jean-Paul Gaudillière, "Rockefeller Strategies for Scientific Medicine: Molecular Machines, Viruses and Vaccines," *Studies in History and Philosophy of Science* 31, no. 3 (2000).

26 John M. Eyler, "De Kruif's Boast: Vaccine Trials and the Construction of a Virus," *Bulletin of the History of Medicine* 80, no. 3 (2006).

27 Caduff, *The Pandemic Perhaps*, 90.

28 Bruno J. Strasser and Thomas Schlich, "A History of the Medical Mask and the Rise of Throwaway Culture," *The Lancet* (2020), https://doi.org/10.1016/S0140–6736(20)31207–1.

29 Lévi-Strauss, *Anthropology and Myth.*

30 Claude Lévi-Strauss, "Language and the Analysis of Social Laws," *American Anthropologist* 53, no. 2 (1951): 158.

17 ELEMENTS-TO-COME

thao phan

This chapter begins with an empty space.

The element that fills this space is classified as "unknown" in the time we call the present, but its conditions of knowability direct us to think of our relation to a time we call the future.

This is the space for the elements-to-come: the elements that we do not yet know but are compelled to keep space for, nonetheless. This is a space defined by its orientation to what is intelligible in the current moment. The elements-to-come asks more than "what might come next?" or "what else could be an element?"; instead, it asks us to consider the distance (or the space) between *what is* and *what could be* called "elemental."

For the chemist Dmitri Mendeleev, it was this empty space that defined his periodic system.[1] His methodical layout of columns and rows was designed not only to identify and catalogue chemical elements and their properties but to demonstrate recurring trends and patterns. It is through this periodic system (or periodicity) that the Table could be used to derive relationships between not only known but as-yet-undiscovered elements. In his earliest iterations of the Table, Mendeleev left gaps within it and speculated on the properties that would logically fill these gaps.[2] Interpolating from the mapped trends, Mendeleev was able to predict the properties of the unknown in impressive detail. His initial predictions were published alongside the original table in 1869 and focused on two gaps in particular: one below aluminium and one below silicon. He named them "eka-aluminum" and "eka-silicon" respectively.[3] By early 1871, he had published detailed descriptions of each. He predicted that eka-aluminium would have an atomic weight of approximately 68, a low melting point, and an oxide character of $5.5 g/cm^3$, and that it would dissolve slowly in both acids and alkalis, among other properties.[4] Similarly, he predicted that eka-silicon would have an atomic weight of approximately 72, a high melting point, and an oxide density of $4.7 g/cm^3$, and that it would be greyish in colour.[5]

Fifteen years later, in 1875, French chemist Emile Lecoq De Boisbaudran, empirically and independent of Mendeleev, identified the element known as gallium. The properties of gallium were almost identical to eka-aluminium down to the method of discovery itself – by means of a spectroscope.[6] Subsequently, in 1886, German chemist Clemens Winkler began analysing an unnamed mineral found at a mine near Freiburg, Saxony. Comparing it to Mendeleev's predictions, and in correspondence with Mendeleev himself, the element would later be identified as eka-silicon and renamed germanium in honour of Winkler's homeland.[7]

From its inception, then, the Periodic Table has been used as a means to wrangle the unknown. Significantly, it does not function to demarcate the limits of knowability but rather seeks to extend its borders. For

Mendeleev, the Table was an efficient template to bring "order and connection"[8]; an opportunity to make a practical contribution to what he
saw as the universal laws of nature. In this way, the Table is an apparatus
that actively shapes the terrain it claims only to describe, extending what
critical decolonial scholars Heather Davis and Zoe Todd call "an idealized version of the world modelled on sameness and replication."[9] It was
this ability to reduce radically different material phenomena to a flat
ontology that Mendeleev saw as the central project of chemistry itself.
Chemistry, he wrote, was

> a natural science which describes homogeneous bodies, studies the mo-
> lecular phenomena by which these bodies undergo transformations into
> new homogeneous bodies, and as an exact science it strives ... to attribute
> weight and measure to all bodies and phenomena, and to recognize the
> exact numerical laws which govern the variety of its studied forms.[10]

The gaps in the Periodic Table were, therefore, not *empty* but *bound to be
filled*, waiting for the right "homogenous body" whose measurements fit
the exact, predetermined criteria.

This predictive logic, based on inference and interpolation, today manifests in the algorithmic infrastructures that are problematically tasked
with managing social life. Predictive analytics and risk models are used
to make determinations on credit and loan eligibility,[11] hireability and
job fitness,[12] exposure to education and housing opportunities,[13] likelihood of crime recidivism,[14] and even whether one is flagged for child
safety or welfare monitoring.[15] The platformization of media – news, mu-
sic, games, film, television, and social networks – enfolds micro aspects
of daily life into regimes of commercial manipulation, driven by the
promise of predictive analytics.[16] On a macro scale, complex simulation
models designed to "predict, prevent, and supress" large scale risks have
birthed new regimes of anticipatory governance.[17] In Australia, amidst
unprecedented national bushfires and the spread of a deadly pandemic,
the language of prediction – future oriented statistics and images of flattened or rising curves[18] – is the primary means by which a government
communicates with a nation caught in cascading and seemingly endless
waves of crisis. Indeed, the year 2020 has, for many, been an object lesson
in how *elemental* predictive logics have become in shaping the affects and
infrastructures of contemporary lifeworlds.

While ostensibly oriented towards the future, in practice many predictive models operationalize the past.[19] In the same way that, as Michelle Murphy writes, "the fullness of our chemical relations" is made
imperceptible by the "pervasive rendering of chemicals as disconnected

functionalist molecules,"[20] human behaviour is ontologically flattened in order to create interoperable data sets. It is not just the case that this data is taken out of context, as anthropologist Nick Seaver has incisively argued, but rather that this data is put to use in an ever-expanding field of contexts.[21] Indeed, within the "nudge economy" all information is useful information if it can potentially trigger desirable (read: profitable) behavioural outcomes.[22] At their worst, these systems not only perpetuate inequality but foreclose alternate possibilities for futures that do not conform to previous patterns or expectations. In this way, predictive systems serve as formalized instruments of racialization and injustice, *elemental* to the forms of racism, sexism, and ableism encoded into "the default settings of technology and society."[23] As many critical race and technology studies scholars have argued, what makes these systems uniquely dangerous is that they travel under the sign of empirically justified objective calculations, making it difficult to enforce accountability or political responsibility.[24] Prediction, here, proceeds on the regressive logic of "sameness and replication" – futures that are intelligible so long as they resemble the past. As with the Periodic Table, these predictions function more like self-fulfilling prophecies that, in Wendy H.K Chun's words, "closes the world it pretends to open."[25]

One of the central goals of this collection has been to disrupt this regressive and cyclical logic. Each chapter represents an effort to intervene into how chemical elements are commonly understood, and in some cases, to stretch the very definition of what counts as elemental in the first place. Where Mendeleev's Table framed elements as indivisible standardized units, the stories collected here illustrate the multiple scales and multiple relations into which every element is nested within. As Timothy Neale argues in his contribution on carbon, creating commensurable units out of relational and indeterminate elements does more than just "make things the same"; it "keeps things the same" by functioning effectively to maintain hegemony.[26]

While also oriented towards a time and place called the future, the elements-to-come is grounded in a wholly different ethico-onto-epistemo-logy to prediction. The elements-to-come is a reconfiguration of "justice-to-come," a phrase most famously associated with the philosophy of Jacques Derrida. In his writing on justice and law, Derrida describes justice as an aporetic experience or "an experience of the impossible."[27] He argues that normative understandings of "a just act" assumes that a correct procedure has been followed, be it judicial, moral, or something else. This procedure is calculable, and defined by "a rule, a norm, or a universal principle."[28] However, what this process describes, he argues, is not justice but law (*droit*). Justice, by contrast, exceeds any determinate rule or law; it demands more than any rule can encompass.[29] He writes,

> Law is the element of calculation, and it is just that there be law, but *justice is incalculable, it requires us to calculate with the incalculable,* and aporetic experiences are the experiences, as improbable as they are necessary, of justice, that is to say of moments in which the decision between just and unjust is never insured by a rule.[30]

Therefore, justice is something that never arrives but is always "to come." Justice, in Derrida's words, is "the experience we are not able to experience."[31]

My use of justice-to-come, however, is not drawn directly from Derrida but comes instead by way of feminist theorist Karen Barad. Unlike the former, Barad's phrase hyphenates each element, linking them as one. In this way, justice-to-come visually illustrates the incalculability of justice, its irreducibility to something that is *knowable in the now,* and instead marks itself as contingent on an indeterminate time we call *to come.* For Barad, justice-to-come, as a single word, serves as an invitation to see justice not as "a state that can be achieved once and for all"[32] but as "an infinite pursuit, an ongoing ethical practice."[33] More than this, justice-to-come invites us to attune to other forms of indeterminacy without surrendering to what Donna Haraway describes as "abstract futurism and its affects of sublime despair and its politics of sublime indifference."[34] It is precisely this ethical attunement that the elements-to-come seeks to draw upon and impress.

Like justice-to-come, elements-to-come is defined by its relation to indeterminacy. Here, the hyphens mark the elements as entities that are never discrete. As the chapters in this collection have demonstrated in vivid detail, elements (chemical or otherwise) cannot be separated from the relationalities they engender nor the stories they unfold. Elements are only intelligible through their "chemical milieus,"[35] in their frictions and circulations in bodies and law,[36] and as experienced through their aftermath and injuries.[37] As Manuel Tironi powerfully states in his contribution on copper, the violence inherent within industrial processes of chemical extraction defy any conceptualization of the elemental as separate to the bodies that co-exist alongside them. While in Chile, mining is said to run in the blood of the nation, it is crucial to specify, in Tironi's words, "whose body does this blood [come] from."[38]

The stories articulated here are quite literally innumerable – too many to be counted. From carbon to copper, mylar to mould, whatever we deem elemental has earned that title through its complex array of entanglements, with bodies (human and non-human, living and non-living), with places (near and far, visible and invisible), with time (still unfolding, already passed, yet to come), and frameworks that help us situate all of these things across vast narratives and scales (the Anthropocene, settler-colonialism, militarism, capitalism, extractivism, and more). Yet, for every elemental story collected here there is an indeterminate number that remain untold, and that we must accept will never be told. Like justice, however, this does not hinder our pursuit but instead figures it as an ongoing ethical practice. In this way, the hyphens that connect the elements-to-come work to eschew the representation of chemicals as structurally isolated, finite beings. They operate as reminders to ensure that even in the practice of writing the word "element" it is never alone on the page, always followed by something that is "to-come."

In addition to this, the elements-to-come asks us to linger on the indeterminacy of the empty space, on the no-thingness, on the gaps within the Table. It asks us to consider how any compilation of elements, from a tabular graph to an edited book, necessarily generates a field of Others. While these Others are not here – not situated under a neat heading, resting between the covers of the book – they are also not *not here*. Indeed, the format of a book or a periodic table, simply cannot capture or contain certain elemental stories, yet we know that they exist and that we, nonetheless, have a responsibility to acknowledge them. The empty space that follows the title "elements-to-come" gestures to the spectre of these innumerable elements.

One might reasonably ask if it is not a contradiction to claim that an empty space contains an infinity of Others. To answer this, I turn to Barad's writing on the void in quantum field theory.[39] Barad explains that in classical physics, there is *literally* nothing emptier than the void. The void is "a spatial frame of reference within and against which motion takes place ... [It] is *that which literally doesn't matter*. It is merely that which frames the absolute."[40] But how do we know the void is empty? Barad suggests that we can set up an experiment that seeks to measure the nothingness of the void, but any apparatus we use to measure would "help constitute and [be] a constitutive part of what is being measured."[41] For instance, if we were to shine a light to attempt to introduce at least one photon (quantum of light) into a vacuum, it would (a) destroy the conditions of that with which we seek to measure, but also (b) capture the entity as either a wave or a particle depending on the kind of apparatus we used to measure it. Barad describes this as an illustration of the quantum principle of ontological indeterminacy: "What we are talking about here is not simply some object *reacting* differently to different probings but *being* differently."[42] This principle of ontological indeterminacy, however, calls the very notion of the zero-energy, zero-matter state of the void into question, as this introduces the possibility of vacuum fluctuations and virtual particles. These fluctuations and particles are not *there* but neither are they not *not there*, or in Barad's words, "The void is not nothing (while also not being something), but rather a desiring orientation toward being/becoming, innumerable imaginings of what might yet be/have been. Nothingness is material (even) in its non/presence."[43]

Like virtual particles and vacuum fluctuations, the elements-to-come points to the indeterminate, mutually constituted excluded Others that haunt the Periodic Table of Elements and this Anthropogenic Table of Elements. That is, that while the elements-to-come marks the final chapter, this does not mean that that the book has necessarily come to an end. The empty space that opens this chapter is not about closure; on the contrary, it represents the "infinite plenitude of openness."[44] Indeed, as Xenia Cherkaev, Heather Paxson, and Stefan Helmreich observe in their chapter, in his later life Mendeleev himself curiously published a piece contradicting his earlier stance on the classification of elements as self-stable units. Comparing these units to a zero he writes "the unit is nothing in and of itself, that it is only a creation of our minds similar to the ones resorted to in geometry when one imagines that a curve is composed of a large number of straight lines."[45] Read in this context, we could also take this to mean that even a seemingly empty zero is comprised of a plenitude of other things. While its shape represents an empty void [0] its curves are comprised of a collection of smaller marks, so many in number they give the impression of a singular, continuous form.

The final point of indeterminacy that I will discuss here (which, of course, is never final) again relates to the principle of ontological indeterminacy. Building on the epistemological framework of quantum physicist Niels Bohr, Barad argues that if an entity can be either a wave or a particle depending on how it is measured, then this suggests that there is no inherent (ontological) distinction that preexists the measurement process.[46] In *Meeting the Universe Halfway*, Barad famously introduced the phrase intra-action to "signify *the mutual constitution of objects and agencies of observation within phenomena* (in contrast to 'interaction,' which assumes the prior existence of distinct entities)."[47] With this in mind, we may ask: how does the material arrangement (in this case a book, based on the Periodic Table) intra-act with the elements within it? Put another way, how does the Anthropogenic Table of Elements, as an apparatus for measuring "what is elemental to this anthropogenic moment,"[48] actively constitute the conditions of intelligibility for what can and cannot be considered elemental? The elements-to-come, in this case, makes explicit the Table as a technology for both knowing and being – an *onto-epistemology*.

In our ongoing experiments with the Table as an apparatus for bringing forth elemental narratives, the editors of this collection staged our own series of workshops. In one iteration at the 4S Annual Meeting in New Orleans,[49] we proposed to "set the table" for an anthropocenic conversation. Punning on the double meaning of the word, we set a literal table, replete with cutlery and tableware. Each guest was invited to "dine" on an element that they had picked from a platter. Using their menus as cues (see figure 0.1) our guests described their elements as an appetizer (what are the "pasts" of your element?), an entrée (how do we live with this element?), and as a dessert (what is the afterlife of the element?). This metaphor of dining immediately invited notions of breaking bread and hospitality. We were nourished on these elemental stories that, ironically, circulated around slow violence and death. The elements on the platter included literal poisons such as sodium monoflouroacetate (also known as the pesticide 1080[50]) and Poly- and perfluorinated Alkyl substances (also known as the firefighting foam PFAS[51]) as well as more benign elements such as aquifers, ice, cement, and sperm. These were stories offered in the spirit of a shared meal, crossing from plate to plate and person to person. This was an onto-epistemology tied to the ethics and norms of hosting and hospitality – an *ethico-onto-epistemological* practice. These experiments have allowed us to surface new questions in relation to the chemical elements. Questions such as: "How do we protect the future from the present? How do we protect the past from its future?"[52]; questions grounded in a sense of responsibility for generations past as well as those to come. By presenting and re-presenting the elements in different forms, we have been able to observe first-hand how each apparatus – a periodic table, a menu, a book collection – intra-acts with each author and each element to iteratively reproduce the meaning of "elemental" anew.

While this essay may conclude the book, it does not point to the end. Like Mendeleev's Periodic Table, to end with the empty space is really to suggest the beginning. The elements-to-come is defined through its indeterminacy. This indeterminacy disrupts the logic of prediction, a logic that by definition seeks to render the future calculable. But the elements-to-come offers a way out of this endless cycle of regressive futures because what is "to come" is always incalculable. What is offered here is nothing less than the infinity of future possibility, the desiring orientation towards something else, something "to come." But rather than rushing to fill the empty space, like Mendeleev, we should instead accept an invitation posed by Barad: "Let us pause before this silence, before rushing on."[53] While it is tempting to seek to fill this space with other elements and elemental stories, or to reconfigure the Table as something other than a book or a conversation game, we should, even just for a moment, pause in recognition of indeterminacy. This might be the indeterminacy of entangled relationalities, the full breadth of which we might not ever know; the indeterminacy of Others, who we can never fully account for; or it might be the indeterminacy of nothingness itself, an empty space.

NOTES

1   Historian Eric Scerri has argued that although there were several periodic systems independently produced in the decade with which Mendeleev's table was published, his "had the greatest impact by far. Not only was Mendeleev's system more complete than the others, but he also worked much harder and longer for its acceptance. He also went much further than the other codiscoverers in publicly demonstrating the validity of his system by *using it to predict the existence of a number of hirtheto unknown elements ...* it was Mendeleev's many successful predictions that were responsible for the widespread acceptance of the periodic system, which his competitors either failed to make predictions or did so in a rather feeble manner." See Eric Scerri, *The Periodic Table: Its Story and Its Significance* (Oxford: Oxford University Press, 2006), 123 (emphasis added).

2   Michael Gordin, *A Well-Ordered Thing: Dmitrii Mendeleev and the Shadow of the Periodic Table*, rev. ed. (Princeton University Press, 2019), 19. See also Addison, this volume.

3   Eka is a prefix based on the Sanskrit name for the digit 1.

4   Scerri, *The Periodic Table*, 133–40.

5   Ibid.

6   The first spectrographic devices had been invented a decade earlier and has led to what historian Michael Gordin describes as a "population explosion in the elemental world." Gordin, *A Well-Ordered Thing*, 19.

7   It should be noted that there were many anomalies that did not match Mendeleev's predictions; for instance, Mendeleev had also speculated that germanium would be difficult to liquify and difficult to volatilize, which prove to be incorrect. Eric Scerri has also noted that Mendeleev has almost a dozen incorrect predictions, which are often overlooked in order to weave a more seamless narrative for Mendeleev's success. Scerri, *The Periodic Table*, 140.

8 Mendeleev "Periodicheskaia zakonnost' khimicheskikh elementov (1871)," quoted in Gordin, *A Well-Ordered Thing*, 30.

9 Heather Davis and Zoe Todd, "On the Importance of a Date, or Decolonizing the Anthropocene," *ACME: An International Journal for Critical Geographies* 16, no. 4 (2017): 761–80.

10 This definition appeared in *The Principles of Chemistry* (*Osnovy khimii*), the text-book in which he published the first version of the Periodic Table. See Mendeleev, "The Principles of Science," quoted in Gordin, *A Well-Ordered Thing*, 21.

11 Tamara K. Nopper, "Digital Character in 'The Scored Society': Fico, Social Networks, and Competing Measurements of Credit Worthiness," in *Captivating Technology: Race, Carceral Technoscience, and Liberatory Imagination in Everyday Life* (Durham: Duke University Press, 2019), 170–87.

12 Winifred R. Poster, "Racialized Surveillance in the Digital Service Economy," in *Captivating Technology: Race, Carceral Technoscience, and Liberatory Imagination in Everyday Life* (Durham: Duke University Press, 2019), 133–68.

13 Julia Angwin and Terry Parris Jr., "Facebook Lets Advertisers Exclude Users by Race," ProPublica, accessed 30 July 2020, https://www.propublica.org/article/facebook-lets-advertisers-exclude-users-by-race.

14 Julia Angwin, Jeff Larson, Surya Mattu, Lauren Kirchner, and ProPublica, "Machine Bias," ProPublica, accessed 30 July 2020, https://www.propublica.org/article/machine-bias-risk-assessments-in-criminal-sentencing?token=sYBNO6t1202JOb6ILFkA_eTWzPmpol3N.

15 Virginia Eubanks, *Automating Inequality: How High-Tech Tools Profile, Police, and Punish the Poor* (St. Martin's Publishing Group, 2018).

16 Mark Andrejevic, "Exploitation in the Data Mine," in *Internet and Surveillance: The Challenges of Web 2.0 and Social Media*, ed. Christian Fuchs (Routledge, 2012), 71–88.

17 Timothy Neale and Daniel May, "Fuzzy Boundaries: Simulation and Expertise in Bushfire Prediction," *Social Studies of Science* 50, no. 6 (13 February 2020): 3.

18 See Tim Rhodes, Kari Lancaster, and Marsha Rosengarten, "A Model Society: Maths, Models and Expertise in Viral Outbreaks," *Critical Public Health* 30, no. 3 (26 May 2020): 253–6.

19 Sun-ha Hong, *Technologies of Speculation: The Limits of Knowledge in a Data-Driven Society* (New York: NYU Press, 2020).

20 Michelle Murphy, "Alterlife and Decolonial Chemical Relations," *Cultural Anthropology* 32, no. 4 (18 November 2017): 496.

21 Nick Seaver, "The Nice Thing about Context Is That Everyone Has It," *Media, Culture & Society* 37, no. 7 (October 2015): 1101–9.

22 Mark Andrejevic, "Automating Surveillance," *Surveillance & Society* 17, no. 1/2 (2019): 7–13.

23 Ruha Benjamin, *Race after Technology: Abolitionist Tools for the New Jim Code* (Medford, MA: Polity, 2019), 98.

24 See Ruha Benjamin, ed., *Captivating Technology* (Durham and London: Duke University Press, 2019); Ruha Benjamin, *Race after Technology: Abolitionist Tools for the New Jim Code* (Medford, MA: Polity, 2019); Virginia Eubanks, *Automating Inequality: How High-Tech Tools Profile, Police, and Punish the Poor* (St. Martin's Publishing Group, 2018); Safiya Noble, *Algorithms of Oppression: How Search Engines Reinforce Racism* (New York: NYU Press, 2018); Sun-ha Hong, *Technologies of Speculation: The Limits of Knowledge in a Data-Driven Society* (New York: NYU Press, 2020).

25 Wendy Hui Kyong Chun, *Queerying Homophily*, 60.

26 Timothy Neale, this volume.

27 Jacques Derrida, "Force of Law: The 'Mystical Foundation of Authority,'" *Cardozo Law Review* 11, no. 5–6 (August 1990): 947.

28 Derrida, *Force of Law*, 949.

29 I owe a great debt of gratitude to my good friend and colleague Tom Sutherland at the University of Lincoln for his generous explication of what is, to me, the "mystical" philosophy of Jacques Derrida.

30 Derrida, *Force of Law*, 947, my emphasis added.

31 Ibid.

32 Karen Barad, *Meeting the Universe Halfway: Quantum Physics and the Entanglement of Matter and Meaning* (Durham and London: Duke University Press, 2007), x.

33 Karen Barad. "After the End of the World: Entangled Nuclear Colonialisms, Matters of Force, and the Material Force of Justice," *Theory & Event* 22, no. 3 (2019): 536.

34 Donna J. Haraway, *Staying with the Trouble: Making Kin in the Chthulucene* (Durham: Duke University Press Books, 2016).

35 See Scott Wark, this volume.

36 See J.R. Latham and Kate Seear, this volume.

37 See Courtney Addison, this volume.

38 Manuel Tironi, this volume.

39 I hesitate to say that I know "nothing" about quantum field theory, because that would then open the infinite possibility that I know something, and so instead I will say that this is an extremely amateur summary of quantum field theory as gleaned through the writing of Barad. I strongly advise going to the source for detailed discussions. Specifically, see Karen Barad, *Meeting the Universe Halfway: Quantum Physics and the Entanglement of Matter and Meaning* (Durham and London: Duke University Press, 2007); Karen Barad, *What Is the Measure of Nothingness? Infinity, Virtuality, Justice, 100 Notes – 100 Thoughts*, Vol. 099 (Kassel, Germany: Documenta, 2012), 6; Karen Barad, "Troubling Time/s and Ecologies of Nothingness: Re-Turning, Re-Membering, and Facing the Incalculable," *New Formations: A Journal of Culture/Theory/Politics* 92, no. 1 (7 April 2018): 77.

40 Karen Barad, "Troubling Time/s and Ecologies of Nothingness: Re-Turning, Re-Membering, and Facing the Incalculable," *New Formations: A Journal of Culture/Theory/Politics* 92, no. 1 (7 April 2018): 77.

41 Karen Barad, *What Is the Measure of Nothingness? Infinity, Virtuality, Justice, 100 Notes – 100 Thoughts*, vol. 099 (Kassel, Germany: Documenta, 2012), 6. See also Barad, *Meeting the Universe Halfway*.

42 Ibid.

43 Karen Barad, *After the End of the World*, 528–9.

44 Karen Barad, *What Is the Measure of Nothingness?*, 17.

45 Xenia Cherkaev, Heather Paxson, and Stefan Helmreich, this volume.

46 Barad, *Meeting the Universe*, 197.

47 Ibid.

48 See the introduction to this collection.

49 This was presented as part of the *Making and Doing* exhibition at the 2019 Annual Meeting of the Society for the Social Studies of Science (4S) in New Orleans. See https://www.4sonline.org/md19/post/a_cordial_invitation _to_the_table_of_elements.

50 See Addison in this collection, as well as Courtney Addison, "Compound 1080 (Sodium Monofluoroacetate)," Theorizing the Contemporary, *Fieldsights*, 27 June 2019, https://culanth.org/fieldsights/compound-1080 -sodium-monofluoroacetate.

51 Timothy Neale, "Poly- and Perfluorinated Alkyl Substances (PFAS)." Theorizing the Contemporary, *Fieldsights*, 27 June 2019. https://culanth.org/fieldsights /poly-and-perfluorinated-alkyl-substances-pfas.

52 Émélie Desrochers-Turgeon, Ozayr Saloojee, and Zoe Todd, this volume.

53 Barad, *Troubling Time/s and Ecologies of Nothingness*, 83.

CONTRIBUTORS

Courtney Addison is a lecturer in the interdisciplinary Centre for Science in Society at Te Herenga Waka, Victoria University of Wellington. Working across anthropology and science technology studies, her doctoral research employed lab and hospital ethnography to explore the everyday ethics of experimental genetic medicine in children. Her current work, an ethnography of 1080 pest control poison, draws together questions of toxicity and belonging to consider what the Anthropocene means in and for Aotearoa.

Brad Bolman is a PhD candidate in the Department of the History of Science at Harvard University. His dissertation, *The Dog Years: Beagling in the Physical and Biological Sciences*, explores the nexus of animal experimentation and capitalism in research with beagle dogs.

Xan Chacko is Mellon Postdoctoral Fellow in Women's and Gender Studies at Wellesley College. Chacko employs a feminist science studies approach to trouble the science of seed preservation and ask what kinds of futures are made possible by the technoscientific endeavour of cryopreservation. Her research interrogates processes of scientific knowledge production to argue for a feminist re-envisioning of credit that is committed to justice. Her current book project traces the rise of cryopreservation as a technique for seed conservation amid changes in environmental policy, intellectual property law, and scientific knowledge. Through an analysis of the material traces and meaning-making practices of scientists and plants, her work explores the history and practices of seed banking and demonstrates how concepts like biodiversity and food security are evoked in a neoliberal era to enable the continuation of extractive colonial practices like plant and seed collecting.

Xenia Cherkaev is a postdoctoral fellow in social anthropology at the Higher School of Economics, St. Petersburg. She holds a PhD in anthropology from Columbia University and is working on two projects: one about the customary use-rights inherent in socialist property law, another about the Soviet and Russian governance of domestic animals. She has written for *The American Historical Review, Cahiers du monde russe, Environmental Humanities, Slavic Review, Bulletin of the Atomic Scientists, Anthropology and Humanism, Ab Imperio, Sotsiologiia Vlasti, Novoe Literaturnoe Obozrenie*, and "Fieldsights" of *Cultural Anthropology*.

Émélie Desrochers-Turgeon is a trained architect and researcher working at the intersections of architectural representation, spatial justice, and landscape. She completed a bachelor's degree in environmental design at Université du Québec à Montréal and a master's degree in architecture at McGill University. Before joining the PhD program, she worked in design firms specializing in industrial design, architecture, and exhibition design in Montreal and Berlin. She is currently a PhD student at the Azrieli School of Architecture & Urbanism. Her doctoral research, funded by the Vanier Canada Graduate Scholarship, examines the encounter between land and architecture, as well as the material culture of the Canadian surveying system as an infrastructure of colonization, utilizing interdisciplinary discourses of architectural representation, settler-colonialism, and landscape.

Eli Elinoff is a senior lecturer in cultural anthropology at Victoria University of Wellington. His research focuses on political and environmental change in urban Thailand. He has published work in *Political and Legal Anthropology Review, The Journal of the Royal Anthropological Institute*, and *Anthropological Theory*. He is the author of *Citizen Designs: Politics and City-Making in Northeastern Thailand* (University of Hawaii Press, 2021) and co-editor of *Disastrous Times: Reconfiguring Environments in Urbanizing Asia* (University of Pennsylvania Press, 2021).

Stefan Helmreich is the Elting E. Morison Professor of Anthropology at the Massachusetts Institute of Technology. His research examines how biologists think through the limits of life as a category of analysis, and he is the author of several books including *Silicon Second Nature: Culturing Artificial Life in a Digital World* (University of California Press, 1998), *Alien*

Ocean: Anthropological Voyages in Microbial Seas (University of California Press, 2009), and *Sounding the Limits of Life: Essays in the Anthropology of Biology and Beyond* (Princeton University Press, 2016).

Frédéric Keck is the director of research at the Laboratory of Social Anthropology (CNRS-Collège de France-EHESS). After working on the history of social anthropology and contemporary biopolitical questions raised by avian influenza, he was the head of the research department of the Musée du quai Branly between 2014 and 2018. He is the author of *Avian Reservoirs: Virus Hunters and Birdwatchers in Chinese Sentinel Posts* (Duke University Press, 2020).

Alison Kenner is an associate professor of politics and a faculty member in the Center for Science, Technology, and Society at Drexel University. Her first book, *Breathtaking: Asthma Care in a Time of Climate Change* (University of Minnesota Press, 2018), documents how care is materialized at different scales to address the US asthma epidemic. From 2014-20, Kenner led the Philadelphia Health and Environment Ethnography Lab (PHEEL), which facilitated collaborative projects between Drexel students, governmental and nongovernmental partners, and community organizations; this included the public education project Climate Ready Philly. Her latest research, *The Energy Rights Project*, looks at how organizations address energy vulnerability in the US mid-Atlantic region.

Janelle Lamoreaux is an assistant professor of anthropology at University of Arizona, conducts social studies of science with emphasis on reproduction and the environment. She has been funded by the National Science Foundation, Social Science Research Council, Wellcome Trust, and Wenner Gren Foundation, and has published in *Cultural Anthropology*, *Cross-Currents: East Asian History and Culture Review*, and various online forums. She is co-editor of the *Handbook of Genomics, Health and Society* (Routledge, 2018). Currently finishing her book on epigenetic toxicology and understandings of male infertility in China, Lamoreaux's next project studies the use of reproductive technologies in coral reef conservation.

J.R. Latham is Alfred Deakin Postdoctoral Research Fellow at Deakin University. He is an interdisciplinary researcher and feminist philosopher whose expertise combines critical concepts of drugs, aging, and narrative with bioethics, queer theory, and science and technology studies (STS) via a focus on gender, sexuality, and medicine. He is also an award-winning writer, recently receiving the Feminist Theory Essay Prize for best article of the year (2017), the Australian Women's and Gender Studies Association Most Distinguished Paper Prize (2018), and the Symonds Prize for excellence in critical inquiry from *Studies in Gender and Sexuality* (2016).

Derek P. McCormack is a professor of cultural geography in the School of Geography and the Environment at the University of Oxford. He has written about nonrepresentational theory, affective atmospheres, and the elements. He is the author of *Refrains for Moving Bodies: Experience and Experiment in Affective Spaces* (2014) and *Atmospheric Things: On the Allure of Elemental Envelopment* (2018), both published with Duke University Press.

Timothy Neale is a DECRA senior research fellow, senior lecturer in anthropology, and convener of the Deakin Science and Society Network at Deakin University. A settler-descendant (Pakeha) from Aotearoa New Zealand, his research concerns the intersections between biopolitics, settler-Indigenous relations, and environmental governance. He is the author of *Wild Articulations: Environmentalism and Indigeneity in Northern Australia* (University of Hawai'i Press, 2017), a producer of the *Conversations in Anthropology* podcast, and an editor of the journal *Science, Technology & Human Values*.

Zeynep Oguz is currently a senior postdoctoral researcher at the University of Lausanne's Social and Cultural Anthropology Lab. Between 2019 and 2021, she was a postdoctoral fellow in environmental humanities at Northwestern University with a joint appointment at the Department of Anthropology. She received her PhD in anthropology in September 2019 at the Graduate Center, the City University of New York. Zeynep's essays and articles on the politics of oil, geology, and coloniality in Turkey have appeared in *Political Geography, Cultural Anthropology, Platypus,* and the *Middle East Report*. She has edited a *Fieldsights* entry on "Geological Anthropology" in *Cultural Anthropology* (2019).

Heather Paxson is the William R. Kenan, Jr. Professor of Anthropology at the Massachusetts Institute of Technology. Her research focuses on how people craft a sense of themselves as moral beings in their everyday lives, especially through activities having to do with family and food. Paxson is the author of *Making Modern Mothers: Ethics and Family Planning in Urban Greece* (University of California Press, 2004) and *The Life of Cheese: Crafting*

Food and Value in America (University of California Press, 2012), and has co-edited the journal *Cultural Anthropology.*

Thao Phan is a research fellow in the ARC Centre for Excellence in Automated Decision-Making and Society and the Emerging Technologies Research Lab at Monash University. She is a feminist technoscience researcher who specializes in the study of gender and race in algorithmic culture. Her work takes an interdisciplinary and intersectional approach, drawing on theory and methods from feminist science and technology studies, media and cultural studies, queer and gender studies, critical race studies, and critical algorithm studies.

Alexis Rider is a PhD candidate in the Department of History and Sociology of Science at the University of Pennsylvania. Her research, which is situated between the history of science, environmental history, and the environmental humanities, explores how ice has been used by naturalists and scientists as a natural chronometer to understand and imagine the deep past and future of the Earth. In addition to her academic writing, Alexis engages in artistic collaborations that explore questions of environmental, particularly cryospheric, change.

Ozayr Saloojee is an associate professor at the Azrieli School of Architecture and Urbanism at Carleton University, a co-director of the Carleton Urban Research Lab, cross-appointed faculty at the University's Institute for African Studies, and associate editor of design for the *Journal of Architectural Education.* Born and raised in Johannesburg, South Africa, he has taught in Canada, Europe, and the US and completed postgraduate degrees at Carleton University and University College London. His research, teaching and academic interests include a focus on politically contested terrains and infrastructure through the intersections of architecture, landscape, cultural geographies, and geo-imaginaries, on sites in Canada, the Great Lakes Basin, South Africa, Turkey, and the Middle East.

Kate Seear is a practising solicitor, associate professor at the Australian Research Centre in Sex, Health and Society at La Trobe University, and an Australian Research Council Future Fellow. Kate has a multidisciplinary background (sociology, gender studies, and the law). Her research is socio-legal and empirical in nature and typically explores connections between law, health, gender, and the body. Her particular interests are in the sociology of law and the ethics of legal practice, the intersections between harm reduction and the law, and alcohol, other drugs, and the law and human rights.

Sarah Stalcup received her master's of science in environmental policy and bachelor's of science in environmental studies and sustainability at Drexel University. Her master's thesis on citizen science explored community-led citizen science and its relation to expert communities through infrastructures such as peer review. She has worked at the Academy of Natural Science in Philadelphia, Pennsylvania, as a curatorial assistant in the Department of Malacology and as a team member of the Patrick Center for Environmental Research's Environmental Policy, Planning and Innovation Team. In addition to her work at the Academy, she has also worked for Fair Tech Collective and the Philadelphia Health and Environment Ethnography Lab.

Manuel Tironi is an associate professor at the Department of Sociology and the Institute for Sustainable Development, both at P. Universidad Católica de Chile. He is also a principal investigator at the Center for Integrated Research on Disaster Risk Reduction. He writes about environmental justice, disasters, politics of care, Indigenous epistemologies, and geological modes of knowing. His research has been published in *Science Technology & Human Values, Social Studies of Science, Geoforum, Sociologial Review, Distinktion,* and *Tapuya,* among others. He currently serves in the editorial collective *Cultural Anthropology.*

Zoe Todd is from Amiskwaciwâskahikan (Edmonton) in Alberta, Canada, and is an associate professor of anthropology in the Department of Sociology and Anthropology at Carleton University. She writes about fish, art, Métis legal traditions, the Anthropocene, extinction, and decolonization in urban and prairie contexts. She also studies human-animal relations, colonialism, and environmental change in north/western Canada. Her research has been published in numerous journals and edited collections including *DIES: Decolonization, Indigeneity, Education, and Society, TOPIA, Journal of Historical Sociology,* and *When the Caribou Do Not Come* (UBC Press, 2018).

Ayo Wahlberg is a professor MSO at the Department of Anthropology, University of Copenhagen. Working within the field of social studies of (bio)medicine, his research has focused on selective reproductive technologies, traditional herbal medicine as well as chronic living. He is the author of *Good Quality: The Routinization of Sperm Banking in China*

(University of California Press, 2018) and co-editor of *Selective Reproduction in the 21st Century* (Palgrave, 2017); he is associate editor at the interdisciplinary journal *BioSocieties*. Ayo is currently wrapping up a five-year project (2015–20) funded by the European Research Council, entitled "The Vitality of Disease – Quality of Life in the Making."

Scott Wark is a research fellow for the project "People Like You: Contemporary Figures of Personalisation," which is funded by a Wellcome Trust Collaborative Award. He is based at the Centre for Interdisciplinary Methodologies at the University of Warwick. He researches online culture, among other things.

INDEX

Cover Image Credits